DAY or AGE

Evaluating Progressive Creationism

David A. McGee

EXEGETICA PUBLISHING
2023

Day or Age: Evaluating Progressive Creationism
©2023 Exegetica Publishing
Published by Exegetica Publishing
        205 S. 8th Street, Suite 310
        Fort Dodge, Iowa 50501
Cover design by Abbie Boyd
ISBN – 978-1-60265-097-8

All Scripture quotations, except those noted otherwise are from the New American Standard Bible, ©1960, 1962, 1963, 1968, 1971, 1972, 1973, 1975, 1977, and 1995 by the Lockman Foundation.

David McGee has written a superb, easy-to-read primer on the errors of Dr. Hugh Ross' progressive day-age creation view, ably demonstrating how it misaligns with orthodox historic Christianity due to his use of scientists' interpretation of natural revelation to override the plain reading of God's special revelation in the Bible. McGee skillfully deals with the main issues at stake, contrasting Ross' mishandling of Scripture and reliance on the secular uniformitarian scientists' claims regarding the age and history of the earth, with the clear plain reading of the biblical text as he defends the literal six-day creation, young-earth position from the historic Christian understanding, science, historic and biblical theology, selected OT and NT texts, and the effects on crucial doctrines of the Christian faith. This concise book is thus a go-to must-read for anyone needing a comprehensive overview of why Ross' progressive day-age creation view should be rejected in favor of the literal six-day, young-earth position advocated by God's Word and historic Christianity.

Andrew Snelling, PhD
Director, Research Division
Answers in Genesis

Throughout Scripture and in other eyewitness events, the Creator has acted in ways that indicate that He is not bound by science. These occurrences leave every Christian at a critical decision point: Do we trust that God exists, independent of science as the Creator of science itself, or do we interpret His Word to fit only within the confines of science, restricting His actions to only explainable scientific knowledge, laws, or occurrences?

Dr. McGee does an excellent job breaking down this "chicken or the egg" situation. The choices the church has been making regarding this dilemma are impacting and permeating many areas of the church, starting with pastoral training and trickling down even to the nursery of the church.

This book is an important work in helping to shape the future of the church, challenging our frail human presuppositions and even breaking down what the church has been teaching us in recent years.

Anthony J. Busti, MD, PharmD, MSc, FNLA, FAHA
CEO at High-Yield Med Reviews
Educator, Physician, Pharmacist, Nurse

In this timely study, McGee explores various Christian interpretations of the earth's age before focusing on Hugh Ross's contribution to the field. Since the subject of the earth's creation affects so many doctrines, McGee's detailed evaluation of the topic according to Scripture's teachings makes this work a vital resource. I highly recommend it!

Matthew Akers, PhD
Associate Dean of Doctoral Programs
Chairman, Old Testament Department
Mid-America Baptist Theological Seminary

One should always be ready to provide a reason for the hope they have within (cf. 1 Pet. 3:15), but when it comes to creation and biblical interpretation, it is all too common for earnest and well-meaning Christians to waffle or throw up their hands when pressed on the issue. One theorist who has had an exceptional amount of influence over post-Darwinian theology is Hugh Ross, and *Day or Age* cuts through a lot of the jargon to clearly summarize and critique Ross's theory. Additionally, McGee addresses, with respect and poise, why evangelicals and biblical inerrantists must wrestle with the issue of creationism because of how it affects the overall interpretation of Scripture, how New Testament writers thought about Genesis 1–11, and the myriad theological concepts hinging on a proper perspective on creationism. *Day or Age* is accessible without being simplistic and makes a strong case for evangelicals to revisit their stance on old-earth creationism, either by intentional thought or by tacit agreement.

Bryce Hantla, EdD
Associate Dean and Associate Professor of Education
Southeastern Baptist Theological Seminary

What a great resource that Dr. David McGee provides in *Day or Age: Evaluating Progressive Creationism*. In it, McGee addresses the significant issues at stake in this debate, including one's view of Bibliology (the doctrine of God's revelation) and Biblical Hermeneutics (principles of interpreting God's revealed Word). McGee provides a convincing case that the Early Church Fathers, theologians of the Middle Ages, and the Reformers, held to a young earth interpretation of Genesis 1–2. Why then the drastic change of course amongst well respected evangelical scholars in many, if not most, Bible colleges and seminaries today? McGee correctly points out that it was **NOT** insights gleaned from the reading of Scripture, but prevailing "scientific" theories introduced in the 18th and 19th centuries by geologists James Hutton, Charles Lyell, and philosopher-scientist, Charles Darwin in their respective writings. McGee identifies the problem, "The ministerial role of observational science usurped Scripture's magisterial role." When instead, as McGee rightly contends, "the primary source of God's revelation, the inerrant Word of God, is magisterial, but nature, a cursed source, is ministerial." In this work, McGee also provides a convincing rebuttal to the arguments of the Progressive Day-Age view, as articulated by its most famous defender, Hugh Ross. If you need a clear and concise book to help you think through the issues, understand the current prevailing views in general, and think carefully about the arguments of Progressive Day-Age Creationism in particular, this is the book for you!

Paul D. Weaver, PhD
Associate Professor of Bible Exposition
Dallas Theological Seminary

In a fair and accurate critique of the works of Hugh Ross (Reasons to Believe), David McGee exposes the cracks in the structure of progressive day-age creationism. In a three-pronged approach, the author first presents a historical survey of Christian interpretations of the Genesis account of creation. What the survey reveals proves to be a continuous and consistent witness to a literal six-day recent creation (not millions and billions of years). Second, Ross's own writings reveal how much secular science governs his approach to the biblical text and minimizes divine revelation from the Creator himself. Third, the contribution and significance of biblical theology demonstrate the priority of Scripture as well as the destructive doctrinal implications of Ross's progressive day-age creationism. Yes, *Day or Age* gifts readers with the equivalent of a seminary course on biblical creationism written for the edification of scholars and laymen alike.

William D. Barrick, ThD
Secretary-Treasurer
Creation Theology Society
Emeritus Professor of Old Testament (retired)
The Master's Seminary

Preface

Does it really matter what one believes about the age of the earth? Aren't we just fighting amongst ourselves when we discuss whether the earth is *billions* or *thousands* of years old? As long as we believe God is the Designer, isn't everything else secondary? These questions come up often in discussions regarding the first chapters of Genesis. My desire to respond biblically to these questions led me to the research that became this book.

The debate about the age of the earth is far from academic. Our understanding of Genesis 1-11 shapes how we understand the authority of Scripture, our place on this planet, and so much more. Origins of life, death, marriage, clothing, disease, and fossils are revealed in the opening chapters of Genesis. Having spent more than a decade investigating these questions, I have found that this issue requires a deep dive. We must, like the Bereans in Acts, examine what others are teaching against the inspired Word of God.

This book will examine the two main views on the Genesis creation account - old earth creationism and young earth creationism. The Bible only supports one of these views. My prayer is that you join me in this examination - test what I wrote against the Word of God - so we can all know what is true, teach what is true, and share what is true.

Contents

Chapter 1

Introduction

PURPOSE OF THE RESEARCH

When it comes to the topic of origins, the evangelical community agrees that Jesus is the creator of the universe. Where this agreement usually ends, however, is on the questions of *how* and *when* Jesus created. Did Jesus begin the creation process billions of years ago or thousands of years ago? Did He create *ex nihilo* (out of nothing), through the evolutionary process of natural selection, or some combination thereof? Whereas some evangelicals are convinced of an older earth (billions of years old) and debate *how* Jesus created it, others maintain the earth is younger (thousands of years old) and affirm a plain-sensed meaning of Genesis 1.[1]

The various interpretations of the Genesis account significantly influence the doctrines of bibliology, anthropology, theology proper, Christology, soteriology, and eschatology. For example, according to old-earth theology, death, disease, and bloodshed existed prior to Adam and Eve's sin.[2] If theologically correct, then the doctrine of the clarity of the Scriptures (bibliology) might be compromised when one considers Genesis 1:31 and

---

[1] The writer has adapted his introduction and overview of progressive day-age creationism and six-day creation from his article. McGee, David. "Critical Analysis of Hugh Ross' Progressive Day-Age Creationism Through the Framework of Young-Earth Creationism." *Answers Research Journal* 12 (2019): 53–71, accessed April 19, 2021, https://answersingenesis.org/creationism/old-earth/critical-analysis-hugh-ross-progressive-day-age-creationism/.

[2] C. John Collins, *Did Adam and Eve Really Exist?* (Wheaton, IL: Crossway, 2011), 116.

Romans 8:18–24. When Jesus declares in Matthew 19:3–6 that God created male and female on the sixth day, yet before their creation, there were supposed billions of years as espoused by old-earth creationists, the character of Jesus (Christology) might be compromised. Lastly, God promises a new heaven and earth without a curse as described in Revelation 22:3. However, according to old-earth creationists, God's newly created world in Genesis 1:1 was filled with death, disease, and bloodshed, and it was declared very good in Genesis 1:31. Old-earth creationists have a different definition of good that, if applied to the new heaven and earth, would include death, disease, and bloodshed. As traditionally interpreted in Revelation 22:3, the eschatological doctrine would be quite different if decay and disease reappeared in the new heaven and earth.

In addition to Christian doctrine, old-earth and young-earth creationists interpret the Bible based upon their general and special revelation presuppositions. Both views assert that the Scriptures are the final authority, yet they have arrived at disparate views regarding the earth's age. One view appears to interpret the creation event by means of general revelation before considering special revelation. Conversely, the other view appears to analyze the creation event first by means of special revelation before considering general revelation. The interpretation of the creation event will reveal whether special or general revelation will be the final authority in determining what God has said through His word. To state it another way, is general revelation magisterial or ministerial to special revelation?

Ernst Mayr, an evolutionary biologist, opines that Christianity's biblical account of creation as told in Genesis 1–2 was "virtually unanimously accepted not only by laypeople but also by scientists and philosophers. The traditional interpretation of the creation account changed overnight, so to speak, in 1859, with the publication of Charles Darwin's *On the Origin of Species.*"[3] However, before Darwin, James Hutton published the *Theory of the*

---

[3] Ernst Mayr, *What Evolution Is* (New York: Basic Books, 2001), 12.

Earth in 1795, and Charles Lyell published the volumes *Principles of Geology* in the 1830s. They sought to dethrone the catastrophism of Noah's flood and replace it with uniformitarianism.

Uniformitarianism is the belief that the universe has always operated the same way in the past as it does in the present. Thus, as described in Genesis, Noah's global flood did not change the earth's topography. Instead, slow, incremental changes occurred geologically. For example, the Colorado River made the Grand Canyon with a little water over millions of years at a constant rate rather than being the result of a deluge of water with sudden catastrophic force.

Evolutionary and old-earth geologists use uniformitarianism as a key to understand the past. Darwin's book popularized Hutton and Lyell's theories by arguing that the God of the Bible was not necessary to explain the universe's origin. The implication was that the creation event in Genesis 1, which indicates the universe is thousands of years old, was not the correct interpretation. Instead, the universe was much older.

The interpretation of Genesis 1 that teaches creation is a recent event can trace its roots to the church fathers and was the prevailing view of Hebrew scholars before the 1860s.[4] Hutton, Lyell, and Darwin challenged the young-earth creation view by suggesting Western society discard the Genesis story and replace it with their "scientific" view, which sought to remove a Creator's necessity. Consequently, the evangelical community, primarily because Darwin popularized the philosophical theory of evolution, has sought to harmonize the scientific interpretations of long periods of time with the early chapters of Genesis to determine the universe's appropriate age.

---

[4] Jeremy Sexton, "Evangelicalism's Search for Chronological Gaps in Genesis 5 and 11: A Historical, Hermeneutical, and Linguistic Critique," *Journal of the Evangelical Society* 61, no. 1 (2018): 5–25; James Mook, "The Church Fathers on Genesis, the Flood, and the Age of the Earth," in *Coming to Grips with Genesis: Biblical Authority and the Age of the Earth*, eds. Terry Mortenson and Thane H. Ury (Green Forest, AR: Master Books, 2008), 23–52.

Because of the influence of Darwin's book, some theologians have sought to reconcile Genesis with the prevailing scientific hypotheses and interpretations.[5] In contrast, other theologians have concluded that no compatibility exists between long periods of time and the Genesis account. One group is old-earth proponents, who believe the universe and earth are billions of years old, and the other group is young-earth proponents, who believe the universe and earth are thousands of years old.

There are four main proposals within the old-earth group. Proposal number one is the *gap theory*, which posits God created the universe as recorded in Genesis 1:1. Afterward, a gap of billions of years elapsed before the present state of creation came into being. Subsequently, in Genesis 1:2, God reformed and refilled the billion-year-old earth in six 24-hour periods. Proposal number two is *theistic evolution*, which affirms the earth is billions of years old but asserts that God used Darwin's mechanism (natural selection) to evolve the flora, fauna, and human beings present on the earth.

Proposal number three is the *framework hypothesis* which seeks to reclassify Genesis 1 as poetic literature rather than historical narrative literature, thus allowing the possibility of billions of years (or whatever the prevailing scientific view of the day happens to be) to fit into the creation event without doing hermeneutical harm. Proposal number four is *progressive day-age creationism* positing that the earth is billions of years old and that each creation day represents many millions or billions of years. In this view, God created but did not use the Darwinian process to evolve the flora, fauna, and human beings.

The effect of proposal number four, progressive day-age creationism (PDAC), chiefly reinforced by Hugh Ross, is prominent within the

---

[5] Kyle Greenwood, *Scripture and Cosmology: Reading the Bible Between the Ancient World and Modern Science* (Downers Grove, IL: IVP Academic, 2015), 212–14. John H. Walton endorsed Greenwood's book. Moreover, Greenwood asserted that Darwin's description of natural selection enhanced the interpretation of the creation account.

evangelical community.[6] For example, Douglas Groothuis, professor of philosophy at Denver Seminary, referenced Ross's book *Creation and Time* in his footnotes before opining: "there is overwhelming evidence that the universe is 13–15 billion years old and that the earth is ancient as well."[7] Norman Geisler, who taught at Dallas Theological Seminary, Trinity Evangelical Divinity School, and founded Southern Evangelical Seminary and Veritas Evangelical Seminary, was quite open to PDAC. He writes, "Not only is it possible that there are time gaps in Genesis 1, but there is also evidence that the 'days' of Genesis are not 6 successive 24-hour days."[8] Immediately after writing his theological assertions, he commended to his readers to *Creation and Time* by Ross. Geisler adds, "It seems plausible the universe is billions of years old . . . there is no demonstrated conflict between Genesis 1–2 and scientific fact . . . a literal interpretation of Genesis is consistent with a universe that is billions of years old."[9] Geisler affirmed PDAC because he believed Ross's assertions on the age of the earth were correct.

Wayne Grudem, the author of *Systematic Theology*, who taught at Trinity Evangelical Seminary and currently teaches at Phoenix Seminary, has affirmed that proposal number four, PDAC, is a valid "option for Christians who believe the Bible today."[10] Interpreting Grudem's view on the age of the earth is confusing. His statement is puzzling because he rejected proposals

[6] Hugh Ross, *Creation and Time: A Biblical and Scientific Perspective on the Creation-Date Controversy* (Colorado Springs: NavPress, 1994), 45. Ross presented his case for his view of progressive day-age creationism in chapter five; however, the thesis of the book is that his view is biblically, theologically, and scientifically correct.

[7] Douglas Groothuis, *Christian Apologetics: A Comprehensive Case for Biblical Faith* (Downers Grove, IL: IVP Academic, 2011), 274.

[8] Norman Geisler, "Does Believing in Inerrancy Require One to Believe in Young Earth Creationism?", accessed April 19, 2021, http://normangeisler.com/does-believing-in-inerrancy-require-one-to-believe-in-young-earth-creationism/.

[9] Norman Geisler, *Systematic Theology*, vol. 2 (Minneapolis: Bethany House, 2003), 650.

[10] Wayne Grudem, *Systematic Theology: An Introduction to Biblical Doctrine* (Grand Rapids: Zondervan, 2000), 297–300.

one and two but accepted proposals three and four and six-day creationism (SDC).

J. P. Moreland at Biola University is open to the possibility of PDAC but not committed. He is less committed to a specific view of old-earth creationism but is certainly against proposal number one. He remarks, "My own views about the creation-evolution controversy are divided between old and young-earth creationism. While I lean heavily toward old-earth views, I do not see the issue as cut-and-dried."[11]

Thus, given the influence of old-earth creationism views in general and the PDAC view specifically within the evangelical community, the purpose of this book is to reveal that Ross's PDAC view, if accepted, will deviate from orthodox Christian theology. Furthermore, the reader will be able to develop a means by which to identify the presuppositions of PDAC and the effect Ross's view has upon established Christian theology.

This book also considers what the church has stated over the past two thousand years regarding the earth's age. Moreover, the book presents research of crucial biblical texts in the Old and New Testaments that contribute to the creation event topic. Those texts established the theological guardrails of an acceptable interpretation of the creation event and then compared it with the PDAC view. The author believes the SDC view best aligns with the authors of the Scriptures and produces a biblical theology consistent with the Bible's storyline. Thus, the book also contrasts the PDAC view with the SDC view.

Although there are other old-earth views, such as proposals number one, two, and three, this book does not discuss these views in excessive detail compared to Ross's view. As established, Ross's theory has had, currently has, and will have a continued influence upon the evangelical community. To analyze his view provides the framework for all other old-earth creation

[11] J. P. Moreland, and John Mark Reynolds, eds., *Three Views on Creation and Evolution* (Grand Rapids: Zondervan, 1999), 142.

views, which seek to justify the prevailing view that the earth is billions of years old.

All old-earth interpretations seek to explain how to interpret Genesis according to the scientific evidence that appears incontrovertible to them. They have revealed a deference to the magisterial role of science over the interpretations of Scripture. In other words, according to old-earth views, the science is settled; therefore, theologians must adjust their interpretation of the Bible. This researcher argues old-earth views are not compatible with the creation event and undermine a foundational doctrine: the authority of the Scriptures.

RELEVANCE OF THE RESEARCH

The influence of Hugh Ross's PDAC view has directly or indirectly swayed a significant number of evangelical seminaries and Christian universities. A thorough understanding of his argument from a theological perspective when compared (or contrasted) with the Scriptures requires investigation. He claims to describe the creation event through the lens of billions of years. Therefore, the average reader should ascertain in the opening chapters of Genesis the location of those billions of years within the text without Ross's aid. If Ross's view is correct, the apostles, who often commented on the Old Testament's writings, should endorse his opinion. Moreover, one should locate within the writings of the early church fathers a view consistent with Ross.

Ross is an astrophysicist and a follower of Jesus. For many evangelical theologians, Ross's firm belief that the universe began billions of years ago settles the earth's age question. As stated previously, numerous evangelicals are promoting old-earth creationism because they rely upon Ross's position that the earth is billions of years old. If his view has any defects, then perhaps evangelical theologians will reconsider their interpretation of the creation event.

Before analyzing Ross's view, this researcher considered some valid questions. For example, does his view affect other doctrinal issues? How does a belief in millions of years of death, disease, and bloodshed prior to Adam's creation affect Paul's words to the Romans (e.g., 5:12 and 8:18–24)? Does Ross's view have any bearing on the context of the gospel? Why did Jesus come to redeem humanity and restore creation if death, disease, and bloodshed (at least in the animal kingdom) were present before Adam and declared very good by God? When Christ returns, will the new heaven and earth replicate the pre-Adamic state and permit death, disease, and bloodshed to continue? What role does observational science have with the Scriptures? Is it magisterial or ministerial? How does a believer determine the correct interpretation of the creation event when there are two diverse views? Can a believer read the Bible with a level of confidence, or must he depend upon scientific observations to determine a passage's correct meaning? What does a belief in the creation event influenced by scientific philosophy mean for the resurrection likelihood? That is, should theologians reconsider the resurrection of Jesus as factual since there is no observational science to support it?

The influence of old-earth creationism is a factor that explains the inconsistent belief of the inerrancy of the Bible among a portion of Southern Baptists.[12] When this author surveyed a sample size of five hundred Southern Baptists in Florida (from approximately one million members), he discovered that evolutionary philosophy had a more substantial influence over old-earth creationists than young-earth creationists. For example, old-earth creationists had a more significant percentage of individuals who doubted the Bible's inerrancy. They also had a more significant percentage who doubted the resurrection of Christ. Moreover, old-earth creationists questioned the authority of the Bible in areas of morality.

---

[12] David A. McGee, "Old Earth Theology: A Factor that Explains Inconsistent Belief of Inerrancy Among Florida Southern Baptists," *Answers Research Journal* 7 (2014): 363–86.

The surveyed Southern Baptists who affirmed young-earth creationism also had been influenced by evolutionary philosophy.[13] Inconsistently, about forty-one percent of them believed there were dinosaur-like creatures on the earth millions of years ago compared to eighty-two percent of old-earth creationists.[14]

The influence of the prevailing scientific view that the earth is billions of years old has persuaded a segment of Southern Baptists to reconsider the early chapters of Genesis. They unashamedly affirm the inerrancy of the Bible, yet a portion of them declare the earth is billions of years old.

Historically, Southern Baptists have addressed the significance of interpreting Genesis on two different occasions: 1876 and 1969.[15] Each time Southern Baptists determined that to deny a literal interpretation of Genesis's creation account, Adam's historicity, and a worldwide Noahic flood eroded inerrancy doctrine. They declared that interpreting the creation event in Genesis is theologically significant and was not a minor theological issue.

Perhaps Ross is correct, and young-earth creationists need to abandon their belief that the Bible affirms a young-earth. If so, they should embrace the current scientific view that the earth is billions of years old. A warranted in-depth look at Ross's ideas compared to the teachings of the Scriptures is warranted. An exegetical analysis will determine if his views match with the plain meaning of the text.

---

[13] An affirmation that the earth was about ten thousand years old or less placed a surveyed Southern Baptist into the category of young-earth creationist.

[14] McGee, "Old Earth Theology," 363–86. One might argue that the affirmation of young-earth creationism would mean no influence of evolutionary philosophy, however, the surveyed young-earth creationists revealed they were inconsistent in their beliefs about the age of the earth.

[15] L. Rush Bush, and Tom Nettles, *Baptists and the Bible* (Nashville: B & H Academic, 1999), 211. See also Randall Williams, "The Role of the Peace Committee in the Southern Baptist Convention Inerrancy Controversy" (PhD diss., Mid-America Baptist Theological Seminary, 2000), 21–23.

MOTIVATION

The motivation to write this book began over twenty-five years ago. The writer has developed projection slides (pre-PowerPoint), published peer-reviewed journal articles, and written a dissertation tangential to this topic.[16] The age of the earth is a secondary issue. What is essential is the authority, reliability, and supremacy of the Bible. The trustworthiness of the opening chapters of Genesis lays a foundation for how one might trust the remaining books of the Bible.

Genesis 1–11, the Bible's opening chapters, are foundational to understanding the central storyline of God's creation, the fall, redemption, and recreation plan. If the reader of Genesis depends on Darwin's popularized teachings and the fluctuating naturalistic philosophy of earth's age, then all the Bible doctrines would frequently be indebted to the future development of humankind's reasoning and observations. To state it another way, the reader can never understand the opening chapters of Genesis because the reader depends on what scientists affirm. Since their views change, so will the interpretation of Genesis. This writer believes the reader can understand the opening chapters of Genesis using the same hermeneutic as is applied to the remaining chapters of Genesis. Evangelical theologians do not generally debate about the historicity of the events in Genesis 12–50. In contrast, they do modify the historicity of the events in Genesis 1–11.

The layperson is left wondering why theologians interpret Genesis 1–11 one way and interpret Genesis 12–50 another way.[17] The believer in the church pew who has come to faith in Christ should have confidence that he can interpret the Bible with proper hermeneutical tools. The believer should

---

[16] David A. McGee, "A Mixed-Methods Study of the Variables that Influenced Florida Southern Baptist' Affirmation of the Inerrancy of the Bible," (EdD diss., Southeastern Baptist Theological Seminary, 2014), 1–264.

[17] Todd S. Beall, "Genesis 1–11: A Plea for Hermeneutical Consistency," *Bible and Spade* 29, no. 2 (2016): 68–74.

also not depend upon the scientific philosophy of the moment to interpret the foundational events of Genesis 1–11.

One potential explanation for the different interpretative methods of Genesis 1–11 and Genesis 12–50 is presuppositions. Old-earth creationism has inadvertently revealed that they affirm in practice that general revelation is magisterial to special revelation. It is a theological blind spot in their methodology. This book explores more deeply the role of general revelation in PDAC and SDC.

Historically, Christian theology has taught the opposite.[18] That is, special revelation is magisterial, while general revelation is ministerial. To put it another way, human scientific observations must agree with the clarity of the Scriptures. This writer believes old-earth creationism elevates general revelation above special revelation, compromising Christian doctrine.

ORGANIZATION OF THE RESEARCH

The writer divides the book into five chapters. Chapter one provides the purpose, relevance, motivation, and methodology of the book. Understanding the significance of the topic is essential to developing the rationale for composing the study.

Chapter two explores the historical overview of Christian interpretations of the earth's age. This section considers the early church fathers, the Middle Ages, and the Reformation period beliefs about the earth's age. A particular emphasis of the chapter explores the eighteenth and nineteenth centuries. During those centuries, the arguments for the development of long periods of time and the assertion that God did not create the world in six twenty-four-hour days began. For the first time, a rival philosophy entered the

---

[18] Millard J. Erickson, *Christian Theology*, 2nd ed. (Grand Rapids: Baker Academic, 1998), 32; Gordon R. Lewis and Bruce A. Demarest, *Integrative Theology*, vol. 1 (Grand Rapids: Zondervan, 1987), 9; Gregg R. Allison, *Historical Theology* (Grand Rapids: Zondervan, 2011), 23.

Western world that did not include Moses's story in Genesis. Darwin popularized long periods of time, but he was not the first to assert the idea. The last section reviews the effect of Morris, Whitcomb, and those young-earth creationists who followed. After Darwinism's rise and the assertion of millions and then billions of years, two men, Morris and Whitcomb, contended that the earth was thousands of years old. They influenced the birth of the modern young-earth creationists' movement that has continued to the present.

Chapter three explores the contribution of Hugh Ross's PDAC view. This researcher espouses and develops the central tenets of Ross. Ross cultivated his arguments through the Hebrew word יוֹם (day), his explanation of day seven of creation, and the inception of death, disease, and bloodshed into the world since the beginning of time. A critical commentary follows that highlights the differences between PDAC and SDC. Since Ross, who has a scientific background, is one of the more popular old-age creationists who has published and given his influence among evangelicals, the author of this book selected his writings as a template for all other old-earth creationists.

Chapter four investigates what the Bible contributes toward the theology of the earth's age. The researcher interprets selected passages in the Law, the Prophets, and the Writings. He also selects passages in the Gospels, the Pauline Epistles, the General Epistles, and John's Apocalypse by reinforcing the canonical interpretation of the creation event. The writer gives special attention to Jesus's words since He claims to be the universe's creator.

Chapter five focuses upon the doctrinal significance of the earth's age based upon the PDAC and SDC views. The author evaluates the doctrines of bibliology, anthropology, theology proper, Christology, soteriology, eschatology, and the approach to apologetics and hermeneutics. A theological reflection upon certain Christian doctrines is worthwhile to see if one's view of the earth's age has any bearing on the Christian faith's fundamentals.

Chapter 2

A Historical Overview
of Christian Interpretations
of the Earth's Age

Historical theology seeks to ascertain how the church has formulated a particular doctrine in the past.[19] This chapter aims to establish what the church historically has affirmed regarding the age of the earth. It also aims to establish what the church historically has considered sound doctrine concerning the age of the earth. Five significant periods will be highlighted: the early church, the Middle Ages, the Reformation, the eighteenth to nineteenth centuries, and the twentieth century.

Establishing what the church has believed historically regarding the earth's age requires recognizing certain presuppositions. Steven Cowan and James Spiegel argued that presuppositions are "truth claims which are assumed without argument."[20] Every truth claim will depend upon arguments, but those arguments will depend upon presuppositions. Eventually, one must cede that presuppositions are foundational to anyone's belief. This author asserts four related to the earth's age.

First, the Holy Spirit has indwelt believers permanently since Pentecost. Embedded into this presupposition is the belief that every generation can accurately interpret the Scriptures through the Holy Spirit's illumination and

---

[19] Gregg Allison, *Historical Theology* (Grand Rapids: Zondervan, 2011), 23.

[20] Steven B. Cowan and James S. Spiegel, *The Love of Wisdom* (Nashville: B & H, 2009), 6.

diligent study. This presupposition does not mean that every generation will accurately interpret every verse in the Bible. However, each generation does have the same Holy Spirit present to guide it. For this reason, the early Christians had the same Holy Spirit as the believers during the Reformation period as well as current-day believers.

The era in which a believer lives does not necessarily hinder or enhance the believer's ability to interpret the text correctly. Access to all the Scriptures in the original languages, the ability to translate the Scriptures, giftedness from the Holy Spirit, and presuppositions based upon hermeneutics will affect one's ability to interpret specific passages of the Bible.[21] God's Spirit was equally active in the second century believers as His Spirit involved Himself with fifteenth century believers. Early Christian writers understood the Scriptures, and their writings (when they interpret accurately) are valuable for succeeding generations.

Second, the body of Christ has existed for almost two thousand years. Since believers are part of Christ's body, there is an interdependence and humility between each succeeding generation. Greg Allison asserted, "the church of today is privileged to enjoy a sense of belonging to the church of the past."[22] Believers of the fifteenth century benefited from commentaries from the second through fourteenth centuries. Believers in the twenty-first century have benefited from the prior works of every era of Christianity. Believers always depend on previous generations of believers.

Providentially, God has used each generation's writings to fine-tune or affirm Christian doctrine. Either they are indebted to the clarity of the doctrine that each generation articulated, or they depend on the affirmation of the doctrine espoused. Humility should be the posture of each generation of believers by recognizing that each generation often depends upon what the previous generation has interpreted.

---

[21] Robert Lightner, *Last Days Handbook* (Eugene, OR: Wipf and Stock, 2005), 143–44.

[22] Allison, *Historical Theology*, 29.

Third, the current generation is not the superior generation. The implicit idea is that the twenty-first century is the sole repository of truth due to scientific advances. To state it another way, one might argue that twenty-first century theologians can see the clearest what the authors of the Bible intended to communicate. According to this view, the advances in science have brought clarity that prior generations did not have.

Scientific advances are self-evident with respect to the invention of penicillin, airplanes, and the cell phone. Historically, Christianity was the soil that cultivated modern science.[23] However, the current-day believer is not entitled to believe scientists have settled scientific knowledge. Some current-day theologians inaccurately conclude that they are the most accurate interpreters of biblical truth because of the interpretations of geologists, biologists, and astrophysicists.

For example, John Collins, a professor of Old Testament at Covenant Theological Seminary in St. Louis, appears to endorse such a view. He does not believe death entered the world because of Adam's sin. He stated, "based on geology and the fossil record, implies that animals had been dying long before human beings came on the scene."[24] Collins rejected the interpretation that Adam's sin ushered in death to the whole world by claiming that current scientific observations are irrefutable. For him, the geological evidence, not the biblical evidence, has settled his interpretation of death before Adam.

To insert current scientific theories into the Scriptures is a dangerous hermeneutic. What one generation considers "settled science" often becomes disputed science in the next generation. For example, Nicolaus Copernicus challenged the egocentricity of the planetary system with a heliocentric theory.[25] Charles Darwin was also confident his view on the species' origin

---

[23] Nancy R. Pearcy and Charles B. Thaxton, *The Soul of Science: Christian Faith and Natural Philosophy* (Wheaton, IL: Crossway, 1994), 17–18.

[24] C. John Collins, *Did Adam and Eve Really Exist?* (Wheaton, IL: Crossway, 2011), 115–16.

[25] Pearcy and Thaxton, *The Soul of Science,* 63–64.

through his natural selection was accurate.[26] Darwin's assertion led him to rule out God as the Creator of any species.[27]

Darwin's confidence escalated and prompted him to assert: "If it could be demonstrated that any complex organ existed, which could not possibly have been formed by numerous, successive, slight modifications, my theory would absolutely break down."[28] Darwin recognized that evidence of an intelligent designer would debunk his view. In the late 1990s and early 2010s, Michael Behe and Stephen Meyer demonstrated that Darwin's view on the species' origin was false biologically and geologically.[29]

At the biological molecular level, Behe examined bacterial flagellum. Bacterial flagellum is irreducibly complex,[30] which means it could not have formed by numerous, successive, slight modifications. Meyer revealed more examples of irreducibly complex species within the geological columns such as Marrella Splendens, Hallucigenia Sparsa, and Opabinia. According to Darwin's theory, species billions of years old should have been simple species.[31] Instead, the geological layers revealed at the Cambrian Era the sudden appearances of fully developed fauna. By his own admission, Darwin's theory has broken down according to his self-defined parameters.

As presented previously, scientific discoveries are in constant flux. Initially, scientists embrace a particular discovery; they dispute the discovery, and sometimes they discard it by suggesting an alternative hypothesis. For this reason, the believer should be wary of relying upon a continuously fluctuating discipline with constant adjustments to interpret God's word. Moreover, believers should be cautious with those within the discipline who seek to remove the Creator from His creation.

---

[26] Charles Darwin, *The Origin of Species* (New York: New American Library, 2003), 9–10.

[27] Darwin, *The Origin of Species*, 7.

[28] Darwin, *The Origin of Species*, 176.

[29] Michael Behe, *Darwin's Black Box* (New York: Free, 2006), 39, 70; Stephen C. Meyer, *Darwin's Doubt* (Nashville: Harper Collins, 2013), 64.

[30] Behe, *Darwin's Black Box*, 39, 70.

[31] Meyer, *Darwin's Doubt*, 29–33.

THE EARLY CHURCH

Genesis explains the origin of all things such as clothes, death, and the seven-day week.[32] The Bible's first book also describes the origin of the universe, plant, animal, human life, marriage, sin, depravity, family, human intelligence, civilization, and language. These seminal topics are significant to believers specifically as well as to humanity generally. What age did the early church fathers ascribe to the earth? Did the early church fathers have an upper limit on the age of the earth? Did the early church fathers believe יוֹם (day) in Genesis 1 was approximately twenty-four hours, an indefinite period of time, or something else? Is there a consensus view regarding the age of the earth that best describes the view of the early church fathers? The following sections reveal what the early church fathers thought on these questions.

Irenaeus (ca. 125–202)

Irenaeus was the Bishop of Lyons.[33] He was a student of Polycarp, who was a disciple of the apostle John. Irenaeus was martyred for his faith. He wrote extensively, but only extant copies of some of the works are available. His theology was formative to the development of the early church.[34] He concluded the creation days of Genesis 1 were approximately twenty-four hours, and the earth was less than six thousand years old:

> And there are some, again, who relegate the death of Adam to the thousandth year; for since "a day of the Lord is as a thousand years," he did not overstep the thousand years, but died within them, thus bearing out the sentence of his sin. Whether, therefore, with respect

---

[32] Ken Ham, *Six Days: The Age of the Earth and the Decline of the Church* (Green Forest, AR: Master Books, 2013), 81.

[33] A. Kenneth Curtis, *Dates with Destiny* (Tarrytown, NY: Fleming H. Revell, 1991), 22.

[34] Alister E. McGrath, *Historical Theology* (England: Wiley-Blackwell, 2013), 38–39.

to disobedience, which is death; whether [we consider] that, on account of that, they were delivered over to death, and made debtors to it; whether with respect to [the fact that on] one and the same day on which they ate they also died (for it is one day of the creation); whether [we regard this point], that, with respect to this cycle of days, they died on the day in which they did also eat.[35]

Irenaeus taught that Adam died before he was one thousand years old. Irenaeus also affirmed that the day Adam sinned was one day, and the duration of the creation day was a regular cycle. The best explanation of Irenaeus's definition of a day is approximately twenty-four hours. Nevertheless, Ross argued that Irenaeus taught each creation day was one thousand years.[36] Irenaeus's commentary on the age of the earth revealed something different:

For in as many days as this world was made, in so many thousand years shall it be concluded. And for this reason the Scripture says: "Thus the heaven and the earth were finished, and all their adornment. And God brought to a conclusion upon the sixth day the works that He had made; and God rested upon the seventh day from all His works." This is an account of the things formerly created . . . and in six days created things were completed: it is evident, therefore, that they will come to an end at the sixth thousand year.[37]

Irenaeus believed God created the world in six days, rested on the seventh day, and that the world will come to an end at six thousand years. He believed the world, during his lifetime, was not billions of years old. Even if one were

---

[35] Irenaeus of Lyons, "Irenaeus against Heresies," *Apostolic Fathers with Justin Martyr and Irenaeus*, eds. Alexander Roberts, James Donaldson, and A. Cleveland Coxe, vol. 1, Ante-Nicene Fathers (Buffalo, NY: Christian Literature Company, 1885), 551–52.

[36] Hugh Ross, *Creation and Time* (Colorado Springs: NavPress, 1994), 18.

[37] Irenaeus, "Irenaeus against Heresies," 551–52.

to grant that Irenaeus's commentary on each creation day's length was unclear, Irenaeus was clear on the maximum age the earth would reach when Christ would return. Ross does not find support from Irenaeus.

Irenaeus also commented that God created each being suited for its environment, and he inferred that each was formed not by a long process but made to function fully in its environment immediately:

> Bestowing harmony on all things, and assigning them their own place, and the beginning of their creation. In this way He conferred on spiritual things a spiritual and invisible nature, on super-celestial things a celestial, on angels an angelical, on animals an animal, on beings that swim a nature suited to the water, and on those that live on the land one fitted for the land—on all, in short, a nature suitable to the character of the life assigned them—while He formed all things that were made by His Word that never wearies.[38]

According to Irenaeus, God created each creation instantaneously, fully developed to thrive in its ecosystem. God did not take millions of years to allow each species to evolve. On the contrary, God designed each species precisely with what it needed on the day He created it.

Justin Martyr (ca. 100–165)

Justin Martyr is considered the greatest apologist of the second century.[39] He lived during a time when the Roman Empire persecuted Christians. Justin Martyr was a defender of the Christian faith during a time when Christians died for their confession that Jesus was Lord. Justin Martyr believed that Plato encountered truth about God through his philosophical maturations of

---

[38] William Edgar and K. Scott Oliphint, eds., *Christian Apologetics Past and Present: A Primary Source Reader to 1500*, vol. 1 (Wheaton, IL: Crossway, 2009), 113.

[39] Edgar and Oliphint, *Christian Apologetics*, 35.

observing nature and logical reasoning. He argued that Plato paved the way for the fullest revelation of God in Jesus. He eventually died for his belief and earned the name Justin Martyr. He remarked,

> Now we have understood that the expression used among these words, "According to the days of the tree [of life] shall be the days of my people; the works of their toil shall abound," obscurely predicts a thousand years. For as Adam was told that in the day he ate of the tree he would die, we know that he did not complete a thousand years. We have perceived, moreover, that the expression, "The day of the Lord is as a thousand years," is connected with this subject.[40]

Similar to Iranaeus, Justin Martyr confirmed that Adam died in less than one thousand years. Justin Martyr appears to teach that Adam's sin happened in one literal day. He also shared the early church fathers' widespread belief that the earth was less than six thousand years based upon the days of creation equaling about one thousand years of real earth history. Safarti explains, "[Church Fathers] believed the world would only last for six thousand years from creation before the return of Christ and the Millennium. In other words, each day of creation corresponded to (but was not equal to) one thousand years of subsequent earth history."[41]

---

[40] Justin Martyr, "Dialogue of Justin with Trypho, a Jew," *Apostolic Fathers with Justin Martyr and Irenaeus*, ed. Alexander Roberts, James Donaldson, and A. Cleveland Coxe, vol. 1, Ante-Nicene Fathers (Buffalo, NY: Christian Literature Company, 1885), 239–40.

[41] Jonathan Sarfati, *Refuting Compromise* (Green Forest, AR: Master Books, 2004), 114.

Basil of Caesarea (329–379)

Basil of Caesarea is also called "Basil the Great." He was a fourth century theologian who wrote on the Trinity.[42] Basil's writings were a precursor of the intelligent design movement in opposition to naturalism.

Ross stressed that Basil is another example of an early church father who had "the difficulty of discovering the date for the universe's creation."[43] He implied that Basil did not focus on when God created the universe; nevertheless, Basil explained his understanding of the creation event:

> *And the evening and the morning were one day.* Why does Scripture say "one day" not "the first day"? Before speaking to us of the second, the third, and the fourth days, would it not have been more natural to call that one the first which began the series? If it therefore says "one day," it is from a wish to determine the measure of day and night, and to combine the time that they contain. Now twenty-four hours fill up the space of one day—we mean of a day and of a night; and if, at the time of the solstices, they have not both an equal length, the time marked by Scripture does not the less circumscribe their duration. It is as though it said: twenty-four hours measure the space of a day, or that, in reality a day is the time that the heavens starting from one point take to return there. Thus, every time that, in the revolution of the sun, evening and morning occupy the world, their periodical succession never exceeds the space of one day.[44]

---

[42] McGrath, *Historical Theology*, 55.

[43] Ross, *Creation and Time*, 21.

[44] Basil of Caesarea, "The Hexaemeron," vol. 8, *St. Basil: Letters and Select Works*, ed. Philip Schaff and Henry Wace, trans. Blomfield Jackson, *A Select Library of the Nicene and Post-Nicene Fathers of the Christian Church, Second Series* (New York: Christian Literature Company, 1895), 64.

Ross has no support from the commentary of Basil to support his view of PDAC. According to Basil, God defined day one by measuring the duration of time that elapses from daylight to nighttime. The length of day one sets the length for the other days. For Basil, each day was twenty-four hours in duration because he described the setting and rising of the sun. There is no doubt that Basil affirmed each creation day as twenty-four hours.

Basil also affirmed that יוֹם (day) in Genesis 1 was not a long period as espoused by Ross. Basil reiterated this point by emphasizing the sun's revolution, the sunrise, and the sunset, which marks the beginning and end of each day:

> God who made the nature of time measured it out and determined it by intervals of days; and, wishing to give it a week as a measure, he ordered the week to revolve from period to period upon itself, to count the movement of time, forming the week of one day revolving seven times upon itself: a proper circle begins and ends with itself... Such is also the character of eternity, to revolve upon itself and to end nowhere. If then the beginning of time is called "one day" rather than "the first day," it is because Scripture wishes to establish its relationship with eternity. It was, in reality, fit and natural to call "one" the day whose character is to be one wholly separated and isolated from all the others. If Scripture speaks to us of many ages, saying everywhere, "age of age, and ages of ages," we do not see it enumerate them as first, second, and third. It follows that we are hereby shown not so much limits, ends and succession of ages, as distinctions between various states and modes of action.[45]

Basil underscored his comprehension of יוֹם (day) in Genesis 1 as twenty-four hours by adding that a week is seven days. He also held that the author

---

[45] Basil, *St. Basil*, 64.

of Genesis did not intend to communicate long, indefinite periods of time. Had the author desired to communicate long periods, he could have used phrases like "age to age" or "ages to ages" attached to each day. Instead, the author of Genesis narrowed the definition of יֹום (day) to mean twenty-four hours.

Gregory of Nissa (ca. 335–395) commented on Basil's writings regarding the days of creation. He believed Basil's words on the creation event are second in importance to the New Testament. Thus, for Gregory of Nissa, Basil's writings were authoritative for the church.

Gregory of Nissa opined, "Before I begin, let me testify that there is nothing contradictory in what the saintly Basil wrote about the creation of the world since no further explanation is needed. They should suffice and alone take second place to the divinely inspired Testament."[46] Gregory of Nissa continued with his understanding of the correct length of each day of creation, adding, "With God's help, we can fathom what the text means, which follows a certain defined order regarding creation. 'In the beginning God created the heavens and the earth' (Gen 1:1), and the rest which pertains to the cosmogenesis which the six days encompass."[47] He affirmed, similar to Basil, that God created the earth in six literal twenty-four-hour days. Considering that Gregory of Nissa equated Basil's words only second to the writings of the New Testament, one can confirm that Gregory of Nissa also believed that God created the earth thousands of years instead of millions of years ago.

Ambrose of Milan (339–397)

Ambrose was the bishop of Milan and was a defender of orthodoxy. He influenced Augustine to convert to Christianity. Ambrose was a man of

---

[46] Gregory of Nyssa, *Hexaemeron*, vol. 44, *Patrologia Graeca*, ed. J. P. Migne, trans. Richard McCambly (Paris: Migne, 1863), 68–69.

[47] Gregory of Nyssa, *Hexaemeron*, 69.

intellect and a fine orator.[48] Ross claimed that Ambrose was not clear on the length of each creation day, thus implying that his PDAC view found support from Ambrose. According to Ross, Ambrose did not fix a date to the age of the earth. However, Ambrose affirmed that יוֹם (day) was twenty-four hours:

> Scripture established a law that twenty-four hours, including both day and night, should be given the name of day only, as if one were to say the length of one day is twenty-four hours in extent . . . Because what is secondary is bound up with what is primary, the nights in this reckoning are considered to be component parts of the days that are counted.[49]

According to Ambrose, Scripture defines the length of each day of creation as twenty-four hours. Ambrose was so confident in the length of each day of creation that he affirmed his belief of twenty-four hours twice. Ambrose also defined a week as seven days and seven nights. Ambrose's writing did not support the PDAC view.

Lactantius (250–325)

Lactantius was a fourth century teacher of rhetoric. The purpose of his writings was to defend Christianity against false religions and heathen philosophy. Christopher Ocker commented that in extreme old age, around 314–317, Lactantius became the tutor of Constantine's ill-fated son, Crispus, whom Constantine executed in 326.[50] As Christianity was growing, Lactantius composed his theological work called *The Divine Institutes*. The

---

[48] Curtis, *Dates with Destiny*, 38–39.

[49] St. Ambrose, "The First Day," vol. 42, *Hexameron, Paradise, Cain and Abel, The Fathers of the Church*, trans. John Savage (Washington, DC: Catholic University of America, 1961), 42–43.

[50] Christopher Ocker, "Lactantius (c. 250–c. 324)," *The Dictionary of Historical Theology* (Carlisle, Cumbria, UK: Paternoster, 2000), 311.

following is what Lactantius wrote in his writings concerning the origins of all things:

> Therefore let the philosophers, who enumerate thousands of ages from the beginning of the world, know that the six thousandth year is not yet completed, and that when this number is completed the consummation must take place . . . God completed the world and this admirable work of nature in the space of six days, as is contained in the secrets of Holy Scripture, and consecrated the seventh day, on which He had rested from His works . . . For there are seven days, by the revolutions of which in order the circles of years are made up. Therefore, since all the works of God were completed in six days, the world must continue in its present state through six ages, that is, six thousand years . . . And again, since God, having finished His works, rested the seventh day and blessed it, at the end of the six thousandth year all wickedness must be abolished from the earth, and righteousness reign for a thousand years; and there must be tranquility and rest from the labours which the world now has long endured.[51]

Lactantius declared that the earth was less than six thousand years old, and God created the world in six days.

Although he did not define םוי (day) as twenty-four hours, he defined each day by revolutions (most likely the observations of the sunset and sunrise). Lactantius shared a similar view with the early church fathers that Christ would return to establish His kingdom at the end of six thousand years

---

[51] Lactantius, "The Divine Institutes," in *Fathers of the Third and Fourth Centuries: Lactantius, Venantius, Asterius, Victorinus, Dionysius, Apostolic Teaching and Constitutions, Homily, and Liturgies*, eds. Alexander Roberts, James Donaldson, and A. Cleveland Coxe, trans. William Fletcher, vol. 7, The Ante-Nicene Fathers (Buffalo, NY: Christian Literature Company, 1886), 211.

of earth history. He did not leave any room for the belief that the earth was millions of years old.

Victorinus of Pettau (250–303)

Victorinus was the Bishop of Pettau and ultimately suffered martyrdom. He wrote commentaries on Genesis, Exodus, Leviticus, Isaiah, Ezekiel, Habakkuk, Ecclesiastes, Song of Songs, Matthew, and Revelation.[52] Much of his original works are lost. However, Jerome preserved some of his writings:

> To me [Victorinus], as I meditate and consider in my mind concerning the creation of this world in which we are kept enclosed, even such is the rapidity of that creation; as is contained in the book of Moses, which he wrote about its creation, and which is called Genesis. God produced that entire mass for the adornment of His majesty in six days; on the seventh to which He consecrated it.
>
> . . . In the beginning God made the light, and divided it in the exact measure of twelve hours by day and by night, for this reason, doubtless, that day might bring over the night as an occasion of rest for men's labours; that, again, day might overcome, and thus that labour might be refreshed with this alternate change of rest, and that repose again might be tempered by the exercise of day.[53]

Victorinus taught that God did not extend each creative act into long, indefinite days. Instead, God created the world quickly. He created the heavens and the earth in six days. Victorinus defined each creation day as two

---

[52] Henry Austin Wilson, "Victorinus (4), ST.," eds. William Smith and Henry Wace, *A Dictionary of Christian Biography, Literature, Sects and Doctrines* (London: John Murray, 1877–1887), 1128.

[53] Victorinus of Pettau, "On the Creation of the World," in *Fathers of the Third and Fourth Centuries: Lactantius, Venantius, Asterius, Victorinus, Dionysius, Apostolic Teaching and Constitutions, Homily, and Liturgies*, eds. Alexander Roberts, James Donaldson, and A. Cleveland Coxe, trans. Robert Ernest Wallis, vol. 7, The Ante-Nicene Fathers (Buffalo, NY: Christian Literature Company, 1886), 341–42.

twelve-hour periods. The daylight part was twelve hours, and the nighttime part was twelve hours. Thus, the creation days were twenty-four hours. He also affirmed the earth was less than six thousand years old.

Ephraim the Syrian (ca. 306–373)

Ephraim (sometimes spelled Ephrem) the Syrian was a theological influence of the early church during the fourth century. He was a deacon, hymn writer, commentary writer, preacher, and theologian.[54] Ephraim wrote a commentary on Genesis and was one of the early church fathers who knew Hebrew. In an example of Ephraim's Hebrew translation relating to the creation account, he interpreted מְרַחֶפֶת in Genesis 1:2 as the Holy Spirit "brooding upon" the waters like a bird on her nest cherishing it, rather than the Holy Spirit being "carried upon" the newly formed creation waters.[55] Ephraim had already begun to contemplate the details of the creation event. He commented on the six days of creation,

> [Moses] then wrote about the work of the six days that were created by means of a Mediator who was of the same nature and equal in skill to the Maker. And after [Moses] said, "This is the book of the generations of heaven and earth," he turned back and recounted those things that he left out and not written about in his first account. . . . In the beginning God created the heavens and the earth, that is, the substance of the heavens and the substance of the earth. So let no one think that there is anything allegorical in the works of the six days. No one can rightly say that the things that pertain to

---

[54] John Gwynn, "Ephraim the Syrian and Aphrahat: Introductory Dissertation," in *Gregory the Great (Part II), Ephraim Syrus, Aphrahat*, eds. Philip Schaff and Henry Wace, vol. 13, *A Select Library of the Nicene and Post-Nicene Fathers of the Christian Church, Second Series* (New York: Christian Literature Company, 1898), 120.

[55] Gwynn, "Ephraim the Syrian," 128.

these days were symbolic, nor can one say that they were meaningless names or that other things were symbolized for us by their names.[56]

Ephraim interpreted the book of Genesis literally and rejected the allegorical interpretation method. He especially emphasized that the six days of creation were not symbolic. Ephraim affirmed that God created the world in six literal days.

Clement of Alexandria (150–215)

Clement of Alexandria was the head of the Catechetical School in Alexandria.[57] He was the first prominent Christian writer from Alexandria. Clement was an apologist who sought to show how secular philosophy depended upon theology. [58] Regarding the days of creation, he asserted that they did not happen over six twenty-four-hour periods. Gerald Bray stated that the days were "rather symbolic expressions of the sequential order of creation."[59] Clement declared in his writings:

> God's resting is not, then, as some conceive, that God ceased from doing. For, being good, if He should ever cease from doing good, then would He cease from being God, which it is sacrilege even to say. . . . For the creations on the different days followed in a most important succession; so that all things brought into existence might have honour from priority, created together in thought, but not being

---

[56] St. Ephrem the Syrian, *Selected Prose Works: Commentary on Genesis, Commentary on Exodus, Homily on Our Lord, Letter to Publius*, ed. Kathleen McVey, trans. Edward G. Matthews, Jr., Joseph P. Amar, Kathleen McVey, vol. 91, *The Fathers of the Church* (Washington, DC: The Catholic University of America Press, 2010), 69, 74.

[57] This writer listed Clement of Alexandria, Origen, and Augustine out of chronological order because Ross believed all three affirmed his view of PDAC with the most substantial quotes.

[58] Gerald Bray, "Clement of Alexandria (c. 150–c. 215)," *The Dictionary of Historical Theology* (Carlisle, Cumbria, UK: Paternoster, 2000), 128.

[59] Mook, *Coming to Grips*, 32.

of equal worth. Nor was the creation of each signified by the voice, inasmuch as the creative work is said to have made them at once. For something must needs have been named first. . . . And how could creation take place in time, seeing time was born along with things which exist.[60]

Although Clement did not describe each creation day as twenty-four hours, he did assert that the creation event happened instantaneously and outside of time. Even though Ross claimed the exact quote as evidence for his view, Clement would reject the PDAC view because a timeless and sudden event cannot last millions of years. [61]

Clement also noted that the solstice was six months, before describing the creation event as instantaneous. It appears that Clement defined a solstice based upon יוֹם (day), to mean approximately twenty-four hours. He wrote, "For the creation of the world was concluded in six days. For the motion of the sun from solstice to solstice is completed in six months—in the course of which, at one time the leaves fall, and at another plants bud and seeds come to maturity."[62]

Origen (182–251)

Origen was born to Christian parents. Septimus Severus, a former Roman Emperor, imprisoned Origen's father and executed him; after that event, Origin had to provide for his family.[63] Eventually, he took to studying and teaching the Scriptures.

---

[60] Clement of Alexandria, "The Stromata, or Miscellanies," in *Fathers of the Second Century: Hermas, Tatian, Athenagoras, Theophilus, and Clement of Alexandria (Entire)*, eds. Alexander Roberts, James Donaldson, and A. Cleveland Coxe, vol. 2, The Ante-Nicene Fathers (Buffalo, NY: Christian Literature Company, 1885), 513.

[61] Ross, *Creation and Time*, 18.

[62] Clement, *Fathers of the Second Century*, 513.

[63] Curtis, *Dates with Destiny*, 26.

He later moved to North Africa and became the president of the school in Alexandria. Origen was a prolific writer and is considered one of the greatest Christian thinkers of the early church fathers.[64] He is mainly responsible for establishing the allegorical method of interpretation. With his method of interpretation, Origen was not sure how to interpret the creation event days. Moreover, he did not believe time had begun during the creation days. For this reason, Ross listed Origen as another example of an early church father who did not affirm a young earth. Origen commented on his understanding of the days of creation and the age of the earth:

> [W]e have treated to the best of our ability in our notes upon Genesis, as well as in the foregoing pages, when we found fault with those who, taking the words in their apparent signification, said that the time of six days was occupied in the creation of the world, and quoted the words: "These are the generations of the heavens and of the earth when they were created, in the day that the LORD God made the earth and the heavens.[65] [Origen adds], After these statements, Celsus, from a secret desire to cast discredit upon the Mosaic account of the creation, which teaches that the world is not yet ten thousand years old, but very much under that, while concealing his wish, intimates his agreement with those who hold that the world is uncreated.[66]

---

[64] James B. Walker, "Origen (c. 185–c. 254)," *The Dictionary of Historical Theology* (Carlisle, Cumbria, UK: Paternoster, 2000), 406.

[65] Origen, "Origen against Celsus," in *Fathers of the Third Century: Tertullian, Part Fourth; Minucius Felix; Commodian; Origen, Parts First and Second*, eds. Alexander Roberts, James Donaldson, and A. Cleveland Coxe, trans. Frederick Crombie, vol. 4, The Ante-Nicene Fathers (Buffalo, NY: Christian Literature Company, 1885), 600–01.

[66] Origen, *Fathers of the Third Century*, 404.

Although Origen is less clear on each creation day's length, he was not an advocate for Ross's view. Origen advocated that the earth was not more than ten thousand years old.

Augustine of Hippo (354–430)

One of the most often cited early church fathers is Augustine. Ross used Augustine's writings to claim that each day of creation was longer than twenty-four hours.[67] Ross added that Augustine's interpretation would allow for progressive day-age creationism: "Augustine took the evenings and morning of the creation days in a figurative sense."[68] Yet, Ross's assertion about Augustine did not support Ross's view. The following commentary reveals that Augustine did not allow for millions of years for each day of creation:

> Let Thy works praise Thee . . . They have therefore their successions of morning and evening, partly hidden, partly apparent; for they were made from nothing by Thee . . . but of concreted matter (that is, matter at the same time created by Thee), because without any interval of time Thou didst form its formlessness . . . Thou hast made the matter indeed of almost nothing, but the form of the world Thou hast formed of formless matter; both, however, at the same time, so that the form should follow the matter with no interval of delay.[69]

Augustine believed God created the six days of creation consecutively while instantaneously. Augustine did not espouse for long periods. He contended

---

[67] Ross, *Creation and Time*, 20.

[68] Ross, *Creation and Time*, 20.

[69] Augustine of Hippo, "The Confessions of St. Augustin," in *The Confessions and Letters of St. Augustin with a Sketch of His Life and Work*, ed. Philip Schaff, trans. J. G. Pilkington, vol. 1, A Select Library of the Nicene and Post-Nicene Fathers of the Christian Church, First Series (Buffalo, NY: Christian Literature Company, 1886), 206.

that the earth was not more than six thousand years old. He concluded about the age of the earth:

> Unbelievers are also deceived by false documents which ascribe to history many thousands of years, although we can calculate from Sacred Scripture that not even six thousand years have passed since the creation of man. It would take too long to show how unfounded all those documents are which chronicle so many thousands of years, and to reveal how unreliable is their authority in this matter.[70]

Augustine ground his belief in the age of the earth from calculations from Scripture. He started with the genealogies in Genesis 5 and 11. In his works, *The City of God*, Augustine provided more commentary concerning the age of the earth:

> According to the letter of Alexander, the kingdom of Assyria lasted more than 5,000 years; whereas, in Greek history it lasted not quite 1,300 years. . . Even then the chronologies of Greek and Egyptian history do not agree; and, since the former [Greek history] does not exceed the true number implied in our Sacred Scripture, it may be accepted.[71]

Augustine rejected Egyptian chronology and argued for the Greek chronology because it fit within the Scriptures' timeframe of the genealogies of Genesis 5 and 11.

In other words, the Scriptures established the age of the earth as no more than six thousand years old. Ross was incorrect in proposing that Augustine

---

[70] Augustine of Hippo, *The City of God, Books VIII–XVI*, ed. Hermigild Dressler, trans. Gerald G. Walsh and Grace Monahan, vol. 14, *The Fathers of the Church* (Washington, DC: The Catholic University of America Press, 1952), 263–64.

[71] Augustine, *The City of God*, 264.

taught an old earth. On the contrary, Augustine affirmed a young earth, an earth not more than six thousand years old and an earth much younger than Ross affirms.

Conclusion of the Early Church Fathers

Irenaeus, Basil of Caesarea, Ambrose of Milan, Victorinus of Pettau, and Ephrem the Syrian asserted that the days of creation were approximately twenty-four hours long. Moreover, Irenaeus, Justin Martyr, Lactantius, Victorinus of Pettau, and Augustine emphasized that the earth was under six thousand years old. Origen concluded that the world was not more than ten thousand years old.

Only Clement of Alexandria avoided affirming that the days of creation were twenty-four hours or that the earth was less than ten thousand years old. Clement did argue that God created instantaneously, and Clement gave no hint that the world was beyond ten thousand years old. Old-earth creationists such as Ross will not find significant support from the early church fathers to affirm their view. On the contrary, the early church fathers emphasized a belief in a recent creation. They were convinced from reading Scripture that God created the earth in six twenty-four-hour days, and the earth was a recent creation.

THE MIDDLE AGES

Aquinas (1225–1274)

Aquinas was the greatest theologian of the Middle Ages.[72] Both university students and professors of theology and philosophy study the words of Aquinas. The catholic church gave him the title of "Doctor Angelicus," which meant he was a giant in the intellectual world.[73] Sproul commented,

---

[72] Timothy S. McDermott, "Aquinas, Thomas (1224/5–74)," *The Dictionary of Historical Theology* (Carlisle, Cumbria, UK: Paternoster, 2000), 25.

[73] R. C. Sproul, *Defending Your Faith* (Wheaton, IL: Crossway Books, 2003), 87.

"The great theologians of history display different styles and different gifts. But for sheer weight of intellect, I doubt that Thomas [Aquinas] has had any peers unless it is the Puritan divine Jonathan Edwards."[74] The following are comments from Aquinas regarding the days of creation:

> The words "one day" are used when day is first instituted, to denote that one day is made up of twenty-four hours. Hence, by mentioning "one," the measure of a natural day is fixed. . . . But the evening and the morning are mentioned as being the ends of the day, since day begins with morning and ends with evening, or because evening denotes the beginning of night, and morning the beginning of day. . . . Thus we find it said at first that "He called the light Day": for the reason that later on a period of twenty-four hours is also called day, where it is said that "there was evening and morning, one day."[75]

Aquinas proclaimed the days of creation were ordinary days of twenty-four hours in length. He affirmed that the phrase "there was evening and morning" added to יוֹם (day) meant a twenty-four-hour period of time. According to Aquinas, each day of creation was the same length, and in six twenty-four-hour periods of time, God created the heavens and the earth.

Conclusion of the Middle Ages

The chief theologian of the Middle Ages was Thomas Aquinas. For Aquinas, the days of creation were six twenty-four-hour periods of time, and he stated his position at two different times. His writings left no doubt that he affirmed a young earth. The writings of Aquinas did not give Ross

---

[74] R. C. Sproul, *The Consequences of Ideas* (Wheaton, IL: Crossway Books, 2000), 65.

[75] Thomas Aquinas, "First Part, Questions 69, 74," in *The Summa Theologica*, trans. Fathers of the English Dominican Province (New York: Benziger Bros., 1947), https://www.ccel.org/ a/aquinas/summa/home.html.

theological support for his PDAC view. Within Ross's writings, one will not find comments from Aquinas regarding the age of the earth.

THE REFORMATION

Martin Luther (1483–1546)

Historical theologians generally identify the beginning of the Reformation with Martin Luther.[76] He posted his famous 95 Theses on the church door in Wittenberg to protest the indulgence abuse. After reading Romans, he concluded that indulgences do not make one righteous. Instead, the one who places his faith alone in Christ alone is made right before God. Luther rediscovered the literal-plain interpretation of Scripture by a close reading of the Bible.[77] He applied this method to his commentary on Genesis.

Luther taught young-earth creationism, believing that God created the earth less than six thousand years ago. He wrote, "From Moses however, we know that 6,000 years ago the world did not exist."[78] Thus, considering when Luther wrote the statement mentioned above, the earth's age is not more than six thousand and five hundred years old.

The following commentary from Luther provides clarity concerning his interpretation of the early chapters of Genesis:

> With respect therefore to this opinion of Augustine, we conclude that Moses spoke literally and plainly and neither allegorically nor figuratively; that is, he means that the world with all creatures was created in six days as he himself expresses it. If we cannot attain unto

---

[76] McGrath, *Historical Theology*, 155.

[77] David S. Yeago, "Luther, Martin (1483–1546)," *The Dictionary of Historical Theology* (Carlisle, Cumbria, UK: Paternoster, 2000), 331.

[78] Martin Luther, "Introduction" in *Commentary on Genesis*, ed. John Nicholas Lenker, trans. Henry Cole (Lutherans In All Lands: Minneapolis: MN, 1904), http://www.gutenberg.org/files/48193/48193-h/48193-h.htm.

a comprehension of the reason why it was so, let us still remain scholars and leave all the preceptorship to the Holy Spirit![79]

All this is moreover intended to lead us into the firm belief and satisfaction of mind that six days were really occupied by God in his creation of all things, contrary to the opinion of Augustine and Hilary, who think that all things were created in a moment.[80] For by the present passage of holy writ our interpretation is confirmed that the six days mentioned by Moses were truly six natural days, because the divine historian here affirms that Adam and Eve were created on the sixth day.[81]

Luther asserted that each day of creation was twenty-four hours in duration. Each creation day was a typical day with an evening and morning. He disagreed with Hilary and Augustine that God created the earth suddenly. Luther then denied that each day was allegorical or figurative. The reason for his interpretation was that the Scriptures confirmed a literal day. Luther's hermeneutic that led him to affirm salvation was based upon faith alone also led him to a young-earth creationist position.

Luther rejected the sudden creation interpretation because the word of God and reason did not permit such an understanding. Only the literal twenty-four-hour day best described the length of the creation days. He remarked, "But with special reference to the sacred passage of Moses before us, how, I pray you, is it possible that six days should be either a moment or

---

[79] Martin Luther, "Introduction" in *Commentary on Genesis*, ed. John Nicholas Lenker, trans. Henry Cole (Minneapolis, MN: Lutherans in All Lands, 1904), http://www.gutenberg.org/files/48193/48193-h/48193-h.htm.

[80] Martin Luther, "Part IV, The Creation of Eve," in *Commentary on Genesis*, ed. John Nicholas Lenker, trans. Henry Cole (Minneapolis, MN: Lutherans In All Lands, 1904), http://www.gutenberg.org/files/48193/48193-h/48193-h.htm.

[81] Martin Luther, "Part VI, God's Work on the Sixth Day," in *Commentary on Genesis*, ed. John Nicholas Lenker, trans. Henry Cole (Minneapolis, MN: Lutherans In All Lands, 1904), http://www.gutenberg.org/files/48193/48193-h/48193-h.htm.

an hour? Neither faith, which rests wholly in the Word, nor reason itself, can admit this."[82]

John Calvin (1509–1564)

One of the titans of the Reformation and the Christian faith is John Calvin. His writings have influenced the evangelical community for almost five hundred years. His most famous work, *The Institutes of the Christian Religion*, is foundational to Reformed and some non-Reformed theology. He was skillful, disciplined, and wrote about the creation days of Genesis.[83] The subsequent commentary of Calvin from four sections of the *Institutes of the Christian Religion* describes his view regarding the creation event:

> Why the world was not created sooner? Answer to it. Shrewd saying of an old man. For the same reason, the world was created, not in an instant, but in six days. The order of creation described, showing that Adam was not created until God had, with infinite goodness, made ample provision for him.[84]

> Here the error of those is manifestly refuted, who maintain that the world was made in a moment. For it is too violent a cavil to contend that Moses distributes the work which God perfected at once into six days, for the mere purpose of conveying instruction. Let us rather conclude that God himself took the space of six days, for the purpose of accommodating his works to the capacity of men.[85]

---

[82] Martin Luther, "Part IV. The Creation of Eve," in *Commentary on Genesis*, ed. John Nicholas Lenker, trans. Henry Cole (Minneapolis, MN: Lutherans in All Lands, 1904), http://www.gutenberg.org/files/48193/48193-h/48193-h.htm.

[83] Curtis, *Dates with Destiny*, 104.

[84] John Calvin and Henry Beveridge, *Institutes of the Christian Religion*, vol. 1 (Edinburgh: The Calvin Translation Society, 1845), 188.

[85] John Calvin, *Commentary on Genesis*, vol. 1, trans. John King (Grand Rapids: Christian Classic Ethereal Library, 1847), 40.

With the same view Moses relates that the work of creation was accomplished not in one moment, but in six days. By this statement we are drawn away from fiction to the one God who thus divided his work into six days, that we may have no reluctance to devote our whole lives to the contemplation of it.[86]

Moreover, as I lately observed, the Lord himself, by the very order of creation, has demonstrated that he created all things for the sake of man. Nor is it unimportant to observe, that he divided the formation of the world into six days, though it had been in no respect more difficult to complete the whole work, in all its parts, in one moment than by a gradual progression.[87]

Calvin contended that God created the world in six consecutive days. He expressed this idea at least four different times in his writing and reinforced his absolute belief. Calvin also rejected the instantaneous creation that Augustine espoused. He contrasted the sudden creation theory with the six-day theory. One can conclude that Calvin affirmed a six-day creation theory that incorporated approximately a twenty-four-hour day. His view about the age of the earth did not align with Ross's PDAC view.

Calvin also commented on his belief in the age of the earth. He wrote, "I admit that profane men lay hold of the subject of predestination to carp, or cavil, or snarl, or scoff. . . . Nor will they abstain from their jeers when told that little more than five thousand years have elapsed since the creation of the world." [88] According to Calvin, even during his day, believing in a young earth had some detractors.

---

[86] John Calvin and Henry Beveridge, *Institutes of the Christian Religion*, vol. 1 (Edinburgh: The Calvin Translation Society, 1845), 188.

[87] Calvin and Beveridge, *Institutes*, 191.

[88] Calvin and Beveridge, *Institutes*, 533.

Similar to Luther, Calvin also affirmed that the earth's age was not much more than five thousand years old. Like his predecessors of the early church fathers who affirmed recent creation, Calvin believed in a recent creation. Calvin recognized that others would mock his belief in a young earth. However, he was confident in his interpretation of the Scripture and that the genealogies of Genesis 5 and 11 led to a recent creation. Like the theologians from the Middle Ages, the Reformers did not give Ross theological support for his PDAC view.

James Ussher (1581–1656)

James Ussher was the archbishop of Armagh.[89] He was in the highest position of the Irish Anglican Church. The result of the Reformation in England produced the Irish Anglican Church. Ussher's most famous work, *The Annals of the World*, is a sixteen hundred page tome of the world's history. He wrote the following about the creation event and age of the earth:

> In the beginning God created the heaven and the earth. The beginning of time, according to our chronology, happened at the start of the evening preceding the 23rd day of October (on the Julian calendar), 4004 BC or 710 JP. (This day was the first Sunday past the autumnal equinox for that year and would have been September 21 on the Gregorian calendar. Historians routinely use the Julian calendar for all BC dates.[90]

Unabashedly, Ussher declared a recent creation of the earth. He calculated day one of the creation as October 23, 4004 BC. Part of his calculation depended upon interpreting Genesis 5 and 11 as historical markers within

[89] F. L. Cross and Elizabeth A. Livingstone, eds., *The Oxford Dictionary of the Christian Church* (Oxford; New York: Oxford University Press, 2005), 1684.
[90] James Ussher, *The Annals of the World* (1658; repr., Green Forest, AR: Master Books, 2006), 17.

the Scriptures to determine the earth's age. He concluded that each creation day was approximately twenty-four hours, and the earth was approximately six thousand years old.

Ross critiqued Ussher's interpretation by arguing that Ussher "ignored Hebrew scholarship and assumed that no generations were omitted from mention in the biblical genealogies [of Genesis 5 and 11]."[91] Ross also argued that Chinese historical records contradicted Ussher's timeline.[92] Moreover, Ross ridiculed Ussher in a children's comic book by depicting Ussher as a buffoon who calculated the earth's age by counting his toes.[93]

Ross concluded that Ussher's attempt to date the earth's age damaged Christians' credibility within the scientific community. A glimpse into Ross's hermeneutical method begins to appear. He affirms the supremacy of Scripture while deferring to other historical documents to critique Ussher. According to Ross, Ussher is misguided because he affirmed the plain meaning of the Genesis creation account and because Ussher did not consult reliable historical documents.

Conclusion of the Reformation

The Reformation's theologians continued interpreting the days of creation and the earth's age exactly like the early church fathers and the theologians of the Middle Ages. They concluded that God created during six consecutive days of an approximate length of twenty-four hours and formed the heavens and the earth about six thousand years ago. Ussher provided an exact year of creation: 4004 BC. Mook commented that theologians had a consensus view on Genesis 1 until the eighteenth century.[94] What changed in the eighteenth century? The answer is that old-earth geologists challenged the traditional interpretation of Moses's writing in Genesis.

---

[91] Ross, *Creation and Time*, 26–27.
[92] Ross, *Creation and Time*, 27.
[93] Sarfati, *Refuting Compromise*, 128.
[94] Mook, *Coming to Grips*, 77.

THE EIGHTEENTH AND NINETEENTH CENTURIES

Preparing the Way for Old-Earth Theology

The beginning of the eighteenth century continued in the tradition of interpreting the days of creation as twenty-four hours and the earth's age about six thousand years. Geology as a science was beginning, and Niels Steenson (Steno) (1638–1686), a geologist, asserted the idea of superposition. The idea of superposition is the hypothesis that the waters settled the debris into sedimentary layers during Noah's deluge. The lower layers were older, and the high layers were younger. The idea of millions of years, however, was foreign to Steno. He affirmed the earth was about six thousand years old.

Terry Mortenson echoed the sentiment regarding the age of the earth by highlighting three geologists from the eighteenth century. He remarked, "English geologists John Woodward (1665–1722), and Alexander Catcott (1725–1779), and the German Geologist Johann Lehmann (1719–1767) wrote books reinforcing this young-earth, global-flood view. Their view was consistent with what the church believed for the first 18 centuries."[95] However, the young-earth, global-flood view changed when geologists James Hutton (1726–1797), Charles Lyell (1797–1875), and philosopher-scientist Charles Darwin (1809–1882) wrote their respective works, which challenged the young-earth view.

The foundation for Hutton's work developed through the writings of Comte de Buffon (1707–1788). Comte de Buffon concluded that the earth was seventy-five thousand years old. He added that the earth needed to be much older, and as scientists extended the age of the earth from thousands to eventually billions, they could discover more truth about the earth.

Unlike the traditional interpretation of Genesis 1:1 that God spoke, Comte de Buffon asserted that the earth began by a comet striking the sun.

---

[95] Terry Mortenson, "'Deep Time' and the Church's Compromise: Historical Background," in *Coming to Grips with Genesis: Biblical Authority and the Age of the Earth*, ed. T. Mortenson and T. H. Ury (Green Forest, AR: Master Books, 2008), 81.

The collision material started the rudimentary development of planets, which eventually led to the earth's formation.[96] In addition to Comte de Buffon, William Whiston (1667–1752), an English theologian, historian, and mathematician, sought to correct the erroneous interpretation that God created the earth in six literal days. Comte de Buffon summarized Whiston:

> This Author [Whiston] commences his treatise by a dissertation on the creation of the world; he says that the account of it given by Moses in the text of Genesis has not been rightly understood; that the translators have confined themselves too much to the letter and superficial views, without attending to nature, reason, and philosophy. The common notion of the world being made in six days, he says is absolutely false, and that the description given by Moses, is not an exact and philosophical narration of the creation and origin of the universe, but only an historical representation of the terrestrial globe.[97]

Comte de Buffon and Whiston wanted to cast doubt upon the writings of Moses in Genesis. Their main reason was that those who had translated the creation event had not considered the scientific, logical, and philosophical arguments in favor of an older earth. They sowed seeds of doubt that Genesis was accurate in describing the origins of the earth. Their comments started clearing the way for a different way of interpreting the origins of the world.

Preparing the way for Darwin's hypothesis of an alternative origin of the species was Jean Lamarck's writings (1744–1829). He was a biologist who

---

[96] Comte de Buffon, *The Epochs of Nature*, eds. and trans. Jan Zalasiewicz, Anne-Sophie Milon, and Mateusz Zalasiewicz (Chicago: University of Chicago, (1778, 2018), xxvi, 40.

[97] Comte de Buffon, "From the System of Whiston" in *Buffon's Natural History*, vol. 1, trans. James Smith Barr (London: Paternoster-Row, (1797, 2014), https://www.gutenberg.org/files/44792/44792-h/44792-h.htm.

suggested that special creation or natural selection could equally account for species variation. Elliot summarized the validity of both options as presented in *Zoological Philosophy* by Lamarck:

> From the à priori view, there does not seem a great deal to choose between the theories; and it was the *à priori* view that was adopted by Lamarck. Many of the known facts of evolution might be accounted for either by use-inheritance or by natural selection. If it is true that acquired characters are inherited, then the giraffe might well have developed his neck through that agency. The hypothesis fits the facts. But so also does the hypothesis of special creation; for if God manufactured the giraffe, neck and all, just as we find him, we immediately reach the goal of our research on the matter. Similarly, again, natural selection is equally satisfactory as an à priori hypothesis.[98]

Lamarck had already developed the embryonic form of evolution. He did not deny that God could have created each specie, while at the same time, he introduced another method called natural selection. Subtly, Lamarck introduced an alternative theory for the origin of the species. No longer was God, as Creator, the only option. An impersonal agent, natural selection, originated by chance and time, could have evolved the species. Lamarck questioned the need for a divine Creator as the only means, the only viable option, for the origin of the world.

James Hutton (1726–1797), from his work, *Theory of the Earth*, asserted that humans might be recent in the earth's history. In contrast, he argued that animals, especially oceanic creatures, are quite ancient. The seeds of old-earth

---

[98] Jean Lamarck, *Zoological Philosophy*, trans. Hugh Elliot (New York: Hafner, 1809, 1914, 1964), xlii, 180, http://www.blc.arizona.edu/ courses/schaffer/449/Lamarck/Lamarck%20Zoological%20Philosophy.pdf.

creationism had begun. Hutton remarked about the origins of humans and animals:

> The Mosaic history places this beginning of man at no great distance; and there has not been found, in natural history, any document by which a high antiquity might be attributed to the human race. But this is not the case with regard to the inferior species of animals, particularly those which inhabit the ocean and its shores. We find, in natural history, monuments which prove that those animals had long existed.[99]

According to Hutton, the oceanic creatures had existed long before humans appeared on the planet. Hutton expanded his theory by implying that God was not necessary. Hutton stated, "we find no vestige of a beginning, no prospect of an end."[100] His opponents accused him of atheism, and Hutton set the stage for Lyell.

After Charles Lyell (1797–1875) wrote his book *Principles of Geology*, the geological community capitulated to an old earth.[101] He argued that understanding the present's geological conditions is the key to understanding the past's geological conditions. For Lyell, the present observation of geological conditions meant an event like Noah's flood did not happen because Lyell had not seen an event like that during his lifetime. For Lyell, observing the present was the key to understanding the past.

According to Lyell, when a geologist locates sedimentary layers, the geologist cannot accept a catastrophic deluge laid down the layers. Thus, the

---

[99] James Hutton, "Chapter 1" in the *Theory of the Earth*, vol. 1 (Royal Society of Edinburgh: Scotland, 1788), https://www.gutenberg.org/files/12861/12861-h/12861-h.htm.

[100] James Hutton, "Chapter 1" in the *Theory of the Earth*, vol. 1 (Royal Society of Edinburgh: Scotland, 1788), https://www.gutenberg.org/files/12861/12861-h/12861-h.htm.

[101] The influential power of scientists has continued to the present. Ross's views, like Lyell's of the past, influence pastors, theologians, and apologists-philosophers.

Noahic deluge did not form the sedimentary layers. Instead, the superposition of materials occurred over a long, indefinite period of time.

Mortenson summarizes, "His [Lyell] theory was a radical uniformitarianism in which he insisted that only present-day processes of geological change at present-day rates of intensity and magnitude should be used to interpret the rock record of past geological activity."[102] Lyell sought to dethrone the Mosaic account of the flood with a non-agency account that eliminated God. Lyell, in his book *Principles of Geology*, explained his reliance upon the uniformity of nature:

> Many appearances, which had for a long time been regarded as indicating mysterious and extraordinary agency, were finally recognized as the necessary result of the laws now governing the material world; and the discovery of this unlooked-for conformity has at length induced some philosophers to infer, that, during the ages contemplated in geology, there has never been any interruption to the agency of the same uniform laws of change. . . . As a belief in the want of conformity in the causes by which the earth's crust has been modified in ancient and modern periods was, for a long time, universally prevalent, and that, too, amongst men who were convinced that the order of nature had been uniform for the last several thousand years, every circumstance which could have influenced their minds and given an undue bias to their opinions deserves particular attention.[103]

Lyell rejected the Noahic deluge as the cause of the sedimentary layers. He sought to undermine Genesis's authority and replace it with the authority of his observations, based upon specific presuppositions. According to Lyell, Noah's flood did not create the geological column. The cause was the

---

[102] Mortenson, *Coming to Grips*, 85.
[103] Charles Lyell, *Principles of Geology* (New York: D. Appleton & Co, 1830), 62–63.

standard, every-day occurrences that had happened over long periods. Thus, there is no special agent, like God, only nature haphazardly laying down sedimentary layers over millions of years.

By the time Darwin (1809–1882) published his book, *The Origin of Species,* a theological shift had already happened. Instead of affirming the straightforward interpretation of Genesis 1–11, a new emphasis among theologians emerged. The new emphasis confirmed the inerrancy of Scripture while seeking to reconcile the "incontrovertible" science with the historic biblically attested interpretation of the days of creation and the earth's age. Some theologians and geologists resisted, [104] but a significant majority capitulated and adjusted their interpretations of Genesis 1–11.

Responses within the Eighteenth and Nineteenth Centuries

Various responses ensued after the popularization of the writings of Hutton, Lyell, and Darwin. A minority affirmed the literal interpretation of Genesis and called themselves "scriptural geologists." Most scriptural geologists lived in England, and few of them lived in America.

However, others responded with capitulation. Unlike the previous seventeen centuries, church theologians in the eighteenth and nineteenth centuries accommodated their previously held views of interpreting Genesis 1–11 to a new theory that accounted for the prevailing scientific views.

Thomas Chalmers (1780–1847), a Scottish minister and professor at both St. Andrews University and the University of Edinburgh, developed a response called the *gap view.* He asserted that God created the universe as recorded in Genesis 1:1. Then he claimed a gap of indefinite time to accommodate the "incontrovertible" science that Lyell proclaimed elapsed before the present state of creation came into being:

---

[104] Terry Mortenson, *The Great Turning Point* (Green Forest, AR: Master Books, 2004), 11, 195–213. See also Richard Tison, "Lords of Creation: American Scriptural Geology and the Lord Brothers' Assault on 'Intellectual Atheism'" (PhD diss., University of Oklahoma, 2008), 383–87.

> The writings of Moses do not fix the antiquity of the globe. If they fix anything at all, it is only the antiquity of the species. . . . Does Moses ever say that there was not a interval [Gap] of many ages betwixt the first act of creation, described in the first verse of the book of Genesis, and said to have been performed at the beginning, and those more detailed operations the account of which commences at the second verse? . . . or does he ever make us to understand that the genealogies of man went any farther than to fix the antiquity of the species, and, of consequences, that they left the antiquity of the globe a free subject for the speculations of the philosophers?[105]

Chambers argued his gap theory view from silence. He also contended that Moses provided hints about the age of the species but not the age of the earth.

According to Chambers, a gap of time existed in the creation narrative: a gap of indefinite time between Genesis 1:1 and Genesis 1:2. Chambers argued this gap existed because God did not deny its existence. After inserting the gap, Chambers believed Genesis 5 and 11 allowed the reader to determine humanity's age, but those same chapters did not reveal the world's antiquity. Hence, for Chambers, humanity was thousands of years old, while the earth was as old as the scientific community declared.

George Stanley Faber (1773–1854), an Anglican theologian, developed a response to the long periods with a view called the *day-age view*. He asserted that each day of creation was an indefinite period of time to accommodate the claims of Lyell. He also denied that Noah's flood was global in scale:

> Thus it appears, that the divine sabbath, instead of being limited to a single natural day, is in truth a period commensurate with the

---

[105] William Hanna, *Memoirs of Thomas Chalmers*, vol. 1 (Edmonston and Douglas: Scotland, 1867), 291.

duration of the creation universe. What that duration will be, no one knows save the Father only: but this we know, that according to the Hebrew chronology the world has already existed nearly six thousand years, and that according to the Samaritan chronology it has existed longer than six thousand years. The divine sabbath therefore is a period of not less duration than six millenaries. . . . Hence, as the seventh day is a period of not less duration than six millenaries, each of the six days must similarly and proportionately have been equivalent to a period equaling or exceeding six thousand years. . . .We have our choice of two expedients. Either a miracle of germination must have been wrought; through which the vegetable seeds sprang up almost instantaneously and quite out of the common course of nature, to complete and productive maturity: or the six days of creation much each have been a period greatly exceeding the length of a single revolution of the earth round its axis. . . . I think, to conclude, that each demiurgic day was a period greatly exceeding the length of a natural day.[106]

Faber proclaimed that because the geologists convinced him that the earth was more than six thousand years old, each creation day must be longer than twenty-four hours.

Faber discovered that the Samaritan chronology was longer than the Hebrew chronology to prove for him that the earth was older. He concluded each day of creation lasted at least six thousand years. He rejected that God created instantaneously on each day what Genesis declared. Instead, the formation of flora and fauna developed over, at minimum, thirty-six thousand years. Each creation day was progressive and extended thousands of years.

---

[106] George Stanley Faber, *A Treatise on the Genius and Object of the Patriarchal, the Levitical, and Christian Dispensations*, vol. 1 (Whitefish, MT: Kessinger, 2009), 116–18.

Charles Spurgeon (1773–1854), the pastor of New Park Street Chapel in London, England, was one of the most prominent preachers in the eighteenth century. His work continues to influence well into the twenty-first century. He accepted the prevailing view that the earth existed long before God created Adam and Eve. He proclaimed in one sermon:

> Can any man tell me when the beginning was? Years ago we thought the beginning of this world was when Adam came upon it, but we have discovered that thousands of years before that, God was preparing chaotic matter to make it a fit abode for man, putting races of creatures upon it, who might die and leave behind the marks of His handiwork and marvelous skill, before He tried His hand on man. But that was not the beginning, for revelation points us to a period long ere this world was fashioned, to the days when the morning stars were begotten when, like drops of dew from the fingers of the morning, stars and constellations fell trickling from the hand of God.[107] Geology tells us that there was death among the various forms of life from the first ages of the globe's history, even when as yet the world was not fitted up as the dwelling of man. This I can believe and still regard death as the result of sin. If it can be proved that there is such an organic unity between man and the lower animals that they would not have died if Adam had not sinned, then I see in those deaths before Adam the antecedent consequences of a sin which was then uncommitted.[108]

Spurgeon believed the world existed long before God created Adam. He also asserted that Scripture points to that ancient world. Spurgeon indicated that

---

[107] Charles Spurgeon, "Unconditional Election," (Sermon, September 2, 1855), 7, https://www.spurgeongems.org/sermon/chs41-42.pdf.

[108] Charles Spurgeon, "Christ the Destroyer," (Sermon, December 17, 1876), 2, https://www.spurgeongems.org/sermon/chs1329.pdf.

geology confirmed death and bloodshed existed before humanity. He was content in his mind to reconcile that God allowed death and bloodshed before Adam, yet Adam's sin was the cause of death. Thus, those species that died before Adam experienced the precursor consequences of sin. In other words, the pre-Adamic world suffered death because Adam would eventually sin against God.

Crawford H. Toy (1836–1919) was a professor of Old Testament at Southern Baptist Seminary. His views on Genesis 1 and 2 conflicted with the seminary's "abstract of principles," and as a result, he resigned. He affirmed that יוֹם in Genesis 1 is a twenty-four-hour day. However, he also denied the gap theory hypothesized by Chalmers. Nevertheless, he denied the accuracy of Genesis 1.

Toy affirmed that what the Bible stated chronologically in Genesis 1, 5, and 11 was wrong scientifically. According to Toy, the Bible was in error.[109] The earth was not thousands of years old; instead, the earth was millions of years old. Ultimately, Toy denied the Bible's inerrancy and full inspiration.[110] Bush and Nettles summarized Toy's theology: "The truth-value of Genesis 1 must be judged by the current scientific theories, which were based on human observations. . . . Toy believed that human observation could distinguish those parts of the Bible that were inspired by God from those parts that were not inspired."[111] Toy believed the scientific community was correct on the age of the earth; therefore, the Bible must be incorrect.

B. B. Warfield (1851–1921) appeared to deny Darwinian evolution yet also affirmed it. That is, he affirmed that the world was ancient, but God did use the evolutionary process. He appeared to conclude that man's age was not much beyond twenty-thousand years yet deferred to the scientific

---

[109] L. Russ Bush and Tom J. Nettles, *Baptist and the Bible* (Nashville, TN: B & H, 1999), 211. Toy was coy with his words. He affirmed that Genesis taught a six-day creation, but the six-day creation event was scientifically untenable.

[110] James Leo Garrett, *Baptist Theology: A Four Century Study* (Macon, GA: Mercer University, 2009), 259–60.

[111] Bush and Nettles, *Baptist and the Bible*, 211.

community to determine the exact age of humanity. Warfield's writings appeared to reflect acquiescence to an old earth with a possibility that humankind was more recent.[112] The writings of Lyell and Darwin affected his views about the creation event. He commented on the genealogies of Genesis 5 and 11 as well as the age of the earth and humanity:

> There is no reason inherent in the nature of the Scriptural genealogies why a genealogy of ten recorded links, as each of those in Genesis [5] and [11] is, may not represent an actual descent of a hundred or a thousand or ten thousand links. The point established by the table is not that these are all the links which intervened between the beginning and the closing names, but that this is the line of descent through which one traces back to or down to the other. . . . And for aught we know instead of twenty generations and some two thousand years measuring the interval between the creation and the birth of Abraham, two hundred generations, and something like twenty thousand years, or even two thousand generations and something like two hundred thousand years may have intervened. In a word, the Scriptural data leave us wholly without guidance in estimating the time which elapsed between the creation of the world and the deluge and between the deluge and the call of Abraham. . . . And it is becoming very generally understood that man cannot have existed on the earth more than some ten thousand to twenty thousand years. . . . The question of the antiquity of man is accordingly a purely scientific one, in which the theologian as such has no concern.[113]

---

[112] Warfield was a leader in the twentieth century movement called Fundamentalism. The movement affirmed five fundamentals which included the inerrancy of Scripture, but the movement did not affirm a recent creation.

[113] B. B. Warfield, "On the Antiquity and Unity of the Human Race," *The Princeton Theological Review* 9, no.1 (1911): 1–25.

Warfield asserted that the genealogies of Genesis 5 and 11 allowed for twenty thousand, two hundred thousand, or even more years. According to Warfield, only science can determine when God created Adam. Warfield also commented on the two options he regarded as viable explanations of humanity's origin. First, God created humanity as described in Genesis. Second, God evolved humanity through natural selection. He elaborated his views:

> It does not appear, however, why this conflict should be pressed to such an extreme. Why should evolutionists insist that the ascent of man must have been accomplished by the blind action of natural forces to the exclusion of all oversight and direction of a high power? Why should biblicists assert that man's creation by the divine fiat must have been immediate in such a sense as to exclude all processes, all interaction of natural forces?[114]

According to Warfield, some form of natural selection via evolution combined with the creation of man, as described in Genesis, was a possible way to explain life's origin. He wanted to combine both views as the fullest explanation. The irony is that natural selection via evolution is a theory devoid of God that undercut Warfield's affirmation of the inerrancy of Scripture ultimately. Warfield had capitulated to the evolutionary theory and saw no conflict with Scripture.

C. I. Scofield (1843–1921) was a pastor, theologian and provided, in 1909, the notes for the Scofield Reference Bible.[115] His notes influenced generations of pastors, laypeople, and theologians of premillennial dispensational-ism. He also taught a form of the gap theory:

---

[114] Mark A. Noll and David N. Livingston, *Evolution, Science, and Scripture: B. B. Warfield, Selected Writings* (Grand Rapids: Baker, 2000), 213.

[115] Scofield was also influential in the foundation of Camino Global (formerly Central American Mission) in 1890.

Scripture gives no data for determining how long ago the universe was created. See notes on Gen. 5:3; 11:10. . . . Two main interpretations have been advanced to explain the expression "without form, and void" (Hebrew *tohu* and *bohu*). The first, which may be called the Original Chaos interpretation, regards these words as a description of an original formless matter in the first stage of the creation of the universe. The second, which may be called the Divine Judgment interpretation, sees in these words a description of the earth only, and that in a condition subsequent to its creation, not as it was originally (see Is. 45:18, note; compare also notes at Is. 14:12; Ezek. 28:12).[116]

Scofield reinforced the gap theory originally espoused by Chambers. For Scofield, Genesis 1:1 stated God created the earth, but something happened to it for an indefinite period of time and ruined the earth. Genesis 1:2 stated God recreated the earth from the original chaotic version, and the second version was where God placed Adam.

Lewis Sperry Chafer (1871–1952) was the founder of Dallas Theological Seminary and a professor of theology. He was a friend of Scofield, influenced by Scofield's dispensationalism, and also appeared to be influenced by Scofield's old-earth creationism:

To make man to be the result of an accidental evolutionary process springing from some supposed primordial germ—which germ itself can not be accounted for apart from a Creator—and all this as a pure imagination fancy without so much as a shadow of substance on which it may rest proof, bears all the marks of mental desperation

---

[116] C. I. Scofield, *Scofield Study Bible* (New York: Oxford University, 1909), 2.

and bankruptcy of ideas. . . theistic evolution . . . is not only un-proved and unreasonable, but is a dishonor to God.[117]

Chafer denied the evolutionary process to explain the origins of humankind. He also disagreed with Ussher's chronology of humanity and believed the Scripture did not conflict with an older earth:

> With respect to his beginning, man is the most recent of all creatures . . . the reasonable extension of human history back several thousand years beyond the dates proposed by Ussher which extension does not conflict, as before stated, with the Biblical record.[118] The phrase *creatio secunda seu mediata* denotes a subsequent act of God by which He brought order and form out of the chaos which followed the original creation.[119]

In his writings, Chafer affirmed humanity's special creation, rejected theistic evolution, yet allowed for an older earth. He did not appear to allow for the age of the earth beyond ninety thousand years. Nevertheless, he was less clear when he described chaos in God's original creation recreated. The writings of Chafer appeared to give deference to Scofield's gap theory.

Conclusion of the Eighteenth and Nineteenth Centuries

Entering the eighteenth and nineteenth centuries, theologians' and scientists' prevailing view regarding Genesis 1 was God created the world over six twenty-four-hour periods. The earth was approximately six thousand

---

[117] Lewis Sperry Chafer, *Systematic Theology*, vol. 2 (Grand Rapids: Kregel, 1993), 130–31. Chafer was not quite as vocal as Scofield regarding the age of the earth, but Chafer saw no conflict with an older earth and the Genesis creation account.

[118] Chafer, *Systematic Theology*, 139.

[119] Lewis Sperry Chafer, *Systematic Theology*, vol. 1 (Grand Rapids: Kregel, 1993), 253.

years old. The predominant young-earth creationist view starting with the early church fathers changed in the eighteenth century.[120] Geologists such as Hutton and Lyell contended that long periods of time elapsed in the geological column. Lyell argued that the key to understanding the past was observing what happens in the present. The hypothesis of uniformitarianism emerged and dethroned catastrophism.

According to Hutton and Lyell, the Noahic flood did not occur as described in the Bible, nor was it a local flood. The result of uniformitarianism and long periods caused theologians, geologists, and pastors to reconsider the historical interpretation of Genesis 1–11.[121] No longer was a literal interpretation espoused from reading the straightforward reading of Genesis 1–11 that affirmed a recent creation; instead, theologians invented ways to reconcile what was convincing evidence in their minds that the earth was much older than six thousand years.[122] They invented the gap theory and progressive day-age creationism to allow long periods and preserve God as the Creator.

An unintentional but inevitable problem followed as a reversal of roles between special and general revelation. The ministerial role of observational science usurped Scripture's magisterial role. Entering the early twentieth century, very few published theologians affirmed a young-earth creation narrative. However, in the middle of the twentieth century, two men challenged the new prevailing view of old-earth creationism from a scientific and scriptural viewpoint and an unlikely publishing partnership.

---

[120] Sarfati, *Refuting Compromise*, 135–37.

[121] Jeremy Sexton, "Evangelicalism's Search for Chronological Gaps in Genesis 5 and 11: A Historical, Hermeneutical, and Linguistic Critique," *Journal of the Evangelical Society* 61, no. 1 (2018): 5–25.

[122] Mortenson, *Coming to Grips*, 101–4.

THE TWENTIETH CENTURY

Liberalism from the previous century continued to challenge orthodox Christianity. A group of conservative scholars and preachers such as G. Campbell Morgan (Westminster Chapel in London), Edgar Y. Mullins (Southern Baptist Seminary), James Orr (United Free Church College in Glasgow), and Benjamin B. Warfield (Princeton Theological Seminary) were concerned about liberalism gaining prominence in the United States and Europe. They developed five fundamentals that they believed all Christians should affirm: the belief in the Bible's inerrancy and infallibility; the affirmation of the virgin birth and deity of Jesus; belief in the substitutionary atonement; the affirmation of the literal physical resurrection of Jesus; and belief in the literal, physical return of Christ.[123] One doctrine they neglected to emphasize was the age of the earth.

Even though the Fundamentalists' movement was orthodox, most leaders affirmed the Lyellian uniformitarianism and the old-earth geological-age system.[124] After the Scopes Trial in 1925, the creation movement, what little remained organizationally, evaporated. Despite the hostile climate toward young-earth creationism, two men who championed the young-earth views from a scientific and biblical viewpoint emerged about twenty-five years later.

Birth of Modern Young-Earth Creationism

In 1961, Henry Morris (1918–2006) and John Whitcomb (1924–2020) coauthored the book *The Genesis Flood*. Morris earned his Doctor of Philosophy in Hydraulics with a minor in Geology, and Whitcomb earned his Doctor of Theology in Hebrew and Old Testament. Both affirmed a

---

[123] David L. Smith, *A Handbook of Contemporary Theology* (Grand Rapids: Victor Books, 1992), 22.

[124] Henry Morris, *History of Modern Creationism*, 2nd ed. (Santee, CA: Institute for Creation Research, 1993), 65.

literal-day creation event and that Noah's flood was global.[125] The book had an almost instantaneous effect. Large numbers of universities, geological societies, and Christian colleges requested the authors to speak.

Morris recalled, behind the scenes, that they had intended to publish with Moody; however, Moody declined because it was well beyond the three-hundred-page limit. Moody also disagreed with their literal-day view and wanted the book to present the other creationists' views. Instead of publishing with Moody, they published with Presbyterian and Reformed Publishing Company.

Modern Twentieth Century Young-Earth Creation Movement

From the publishing of *The Genesis Flood*, the formation of four influential young-earth creationist groups started. First was the Creation Research Society, which began in 1963 and still exists.[126] They are a professional organization of trained scientists and interested laypersons firmly committed to scientific, special creation.

The second was the Institute of Creation Research (ICR), which began in 1970 under Morris' leadership.[127] They are a professional organization of scientists and theologians. Until the early 2000s, ICR was the most influential young-earth creationist organization in the United States.

Third was Answers in Genesis (AIG), which began in 1977 in Australia, initially call the Creation Science Foundation.[128] AIG has become the most popular young-earth creationist organization in the United States. It opened a museum dedicated to creation studies and reproduced an identical size

---

[125] John C. Whitcomb and Henry M. Morris, *The Genesis Flood: The Biblical Record and Its Scientific Implications* (Phillipsburg, NJ: Presbyterian and Reformed Publishing, 1961), 1–2.

[126] Morris, *History of Modern Creationism*, 195.

[127] Morris, *History of Modern Creationism*, 271.

[128] J. B. Stump, ed., *Four Views on Creation, Evolution, and Intelligent Design* (Grand Rapids: Zondervan, 2017), 14.

replica of Noah's ark with the exact dimensions. AIG has offices in Australia, Canada, the United Kingdom, Mexico, Peru, and the United States.

The fourth was the Creation Ministries International, which began in 2006[129] in Australia and has since opened offices in Canada, New Zealand, Singapore, South Africa, the United Kingdom, and the United States.

Since the publication of *The Genesis Flood*, a growing but still minority of colleges and seminaries affirm young-earth creationism within the United States.[130] The majority of Christian colleges and seminaries do not officially endorse six-day creationism. For example, all six Southern Baptist Convention Seminaries have not officially affirmed young-earth creationism.

The outcome has been that most theologians, Christian apologists, and Old Testament scholars within the evangelical community affirm some form of old-earth creationism. As mentioned in chapter one, Douglas Groothuis (1957–Present), Norman Geisler (1932–2019), J. P. Moreland (1948–Present), and Wayne Grudem (1948–Present) are examples of prevalent old-earth creationism. Nevertheless, existing presently are pastors, apologists, and scientists who are young-earth creationists and are producing literature. Some examples are John F. MacArthur (1939–Present), Ken Ham (1951–Present), Kurt Wise (1959–Present), and William D. Barrick (1946–Present).[131]

---

[129] Morris, *History of Modern Creationism*, 347.

[130] A few prominent examples are Cedarville University in Ohio, The Master's College and Seminary in California, San Diego Christian College in California, Bryan College in Tennessee, and Grace College and Seminary in Indiana. Many of the largest Christian universities, such as BIOLA University in California, Liberty University in Virginia, California Baptist University in California, Grand Canyon University in Arizona, Wheaton College in Illinois, and Calvin College in Michigan, have not made a public endorsement. Answers in Genesis has a list of colleges and seminaries who have affirmed six-day creationism and the earth is not beyond ten thousand years old, accessed April 19, 2021, https://answersingenesis.org/colleges /colleges-and-universities.

[131] John MacArthur, "Forward," in *Coming to Grips with Genesis: Biblical Authority and the Age of the Earth*, eds. T. Mortenson and T. H. Ury (Green Forest, AR: Master Books, 2008), 9–13; Ken Ham, *Six Days: The Age of the Earth and the Decline of the Church* (Green Forest, AR: Master Books, 2013), 81; Kurt Wise, *Faith, Form, and Time: What the Bible*

Summary of the Twentieth Century

The majority view of theologians at the start of the twentieth century embraced some form of old-earth creationism. They generally rejected Darwinian evolution yet affirmed Lyellian uniformitarianism and the old-earth geological-age system. When Morris and Whitcomb published the book, *The Genesis Flood,* a revitalization of young-earth creationism began. Because of their book, a new generation of young-earth creationist organizations launched. The most influential creationist organizations in the United States are AIG and ICR. However, most Christian colleges, universities, and seminaries like the six Southern Baptist Seminaries and Wheaton University have not officially endorsed young-earth creationism. A significant minority, like Cedarville and The Master's University, have endorsed a belief in a six-day twenty-four-hour creation week and a universe less than ten thousand years old.

SUMMARY OF HISTORICAL INTERPRETATIONS
OF THE EARTH'S AGE

From the early church fathers to the eighteenth century, the same consensus view interpreted the creation event in Genesis 1. The belief was that God created the earth in six days. Each day was approximately twenty-four hours. The universe's age was approximately six thousand years. However, a few theologians placed an upper limit of ten thousand years. Theologians like Origen and Augustine taught that God created instantaneously rather than during six twenty-four periods. They still affirmed that the universe was less than 10,000 years old.

Teaches and Science Confirms about Creation and the Age of the Universe (Nashville: B & H, 2002), 44–45; William D. Barrick, "Old Testament Evidence for a Literal, Historical Adam and Eve,"in Searching for Adam: Genesis & the Truth About Man's Origin, ed. Terry Mortenson (Greenfield, AR: Master Books, 2016), 17-52.

Hutton, Lyell, and Darwin popularized uniformitarianism, which led to an old-earth geological system. The Christian community responded by rejecting, for the most part, Darwinian biology, but they accepted Lyellian uniformitarianism. A few theologians in the United Kingdom and the United States, called scriptural geologists, resisted the old-earth creationism and clung to the Scriptures to defend young-earth creationism. Nevertheless, in the next one-hundred and fifty years, old-earth creationism was the dominant view of evangelicalism and their churches.

A change happened in the middle of the twentieth century when Morris and Whitcomb wrote *The Genesis Flood*. Their book recaptured the historical interpretation of Genesis 1. Morris and Whitcomb affirmed a young-earth and the literality of the Noahic deluge. The influence of their books stirred theologians, pastors, and scientists to form young-earth creationist societies. The two most influential in the United States are AIG and ICR. AIG and ICR built creation museums, and AIG constructed an identical size replica of Noah's Ark. The effect of Morris and Whitcomb's book stirred some theologians to reconsider Genesis 1–11 from a young-earth creationist interpretation.

Currently, some seminaries and Christian colleges endorse young-earth creationism; however, they are still in the minority. The majority of evangelical seminaries and Christian colleges are either against young-earth creationism or silent. The old-earth views that Hutton, Lyell, and Darwin asserted in the eighteenth and nineteenth centuries have continued into the twenty-first century.

Chapter 3

Analysis and Evaluation of Hugh Ross's Progressive Day-Age Creationism

The most vocal proponent of progressive day-age creationism (PDAC) is Hugh Ross and those connected with the ministry Reasons To Believe (RTB). Ross earned his Doctor of Philosophy in Astronomy from the University of Toronto and founded RTB in 1986. He has written dozens of books and articles on this topic and, most recently, in 2017, was one of four contributors to the book *Four Views on Creation, Evolution, and Intelligent Design*. Ross has made many appearances on media outlets and written numerous news articles. Hugh Ross best represents PDAC, which teaches that "evidence of a cosmic beginning in the finite past–only 13.8 billion years ago" agrees with Genesis 1.[132] Ross also asserted that the days of creation in Genesis 1 must be long definite periods of time.[133]

When Ross published *Creation and Time*, his book received endorsements from Walter Kaiser (Gordon-Conwell Seminary), Earl Radmacher (Western Seminary), Stan Oakes, Ted Martin (Campus Crusade), and Jim Berney (Intervarsity).[134] As previously stated in chapter one, Hugh Ross has directly influenced a significant number of evangelical scholars and indirectly

---

[132] Hugh Ross, *Navigating Genesis: A Scientist's Journey through Genesis 1–11* (Covina, CA: RTB, 2014), 15.

[133] Hugh Ross, *Creation and Time* (Colorado Springs: NavPress, 1994), 45, 101.

[134] Ross, *Creation and Time*, 1.

influenced Christian colleges and seminaries. For example, Douglas Groothuis at Denver Seminary referred readers to Ross's writings.[135] Grudem currently at Phoenix Seminary and previously at Trinity Evangelical Divinity School stated that progressive day-age creationism (PDAC) is a valid way to interpret the Genesis creation event.[136]

In their work, *Integrative Theology*, Gordon Lewis and Bruce Demarest summarized their belief that the most reasonable understanding of יוֹם (day) in Genesis 1 aligns with PDAC. They remarked, "The most probable conclusion is that the six consecutive creative acts were separated by long periods of time."[137] Lewis and Demarest added their doubts about the geological viability of Noah's flood by explaining the immense discovery of fossils: "The vast majority of professionally engaged geologists, both Christian and non-Christian, reject the arguments for flood geology as indefensible science."[138] Although Lewis and Demarest do not use Ross's name, his viewpoint shines through their writings. Like Ross, they disclosed their presupposition that the scientific evidence warrants adjusting the historical, plain-sensed meaning interpretation of Genesis 1.

Reputable evangelical theologians promote old-earth creationism in part because they rely upon Ross's confidence that the earth is billions of years old. His influence warrants an in-depth look at his ideas compared to the teachings of the Scriptures. A critical analysis of his tenets will determine if his views agree with the biblical text's plain meaning.

---

[135] Douglas Groothuis, *Christian Apologetics: A Comprehensive Case for Biblical Faith* (Downers Grove, IL: IVP Academic, 2011), 274.

[136] Wayne Grudem, *Systematic Theology: An Introduction to Biblical Doctrine* (Grand Rapids: Zondervan, 2000), 297–300.

[137] Gordon R. Lewis and Bruce A. Demarest, *Integrative Theology*, vol. 2 (Grand Rapids: Zondervan, 1990), 44.

[138] Lewis and Demarest, *Integrative Theology*, 46.

THE CENTRAL TENETS OF HUGH ROSS'S
DAY-AGE CREATIONISM

Ross affirmed five central tenets that led him to conclude that the earth is billions of years old. First, there are two sources of revelation. Second, the Hebrew word יוֹם, translated as *day* in English, does not mean a twenty-four-hour day in Genesis 1. Third, the unending seventh day of the creation week allows for interpreting a long period of time. Fourth, Noah's flood was universal but not global. Fifth, Ross presented a different backdrop for the gospel. According to Ross, the sin of Adam did not bring about nonhuman death, disease, and bloodshed.[139] This chapter will consider these five central tenants.

Two Sources of Revelation

The foundational premise of PDAC is its view of two sources of revelation. Ross affirmed that two inerrant sources justify PDAC. Those two sources are the Bible and nature:

> Some readers might fear that I am implying that God's revelation through nature is somehow on an equal footing with His revelation through the words of the Bible. Let me simply state that truth, by definition, is information that is perfectly free from contradiction and error. Just as it is absurd to speak of some entity as more perfect than another, so also one revelation of God's truth cannot be held as inferior or superior to another.[140]

---

[139] Ross, *Creation and Time*, 60–61. This author would add that according to the New Testament commentary on Genesis 3:14–19, death, disease, and bloodshed of nonhuman life began with the sin of Adam (Rom. 8:18–23).

[140] Ross, *Creation and Time*, 57.

Ross recognized the fear that readers might have after reading this quote to mean that he had elevated nature to be equal with the Scriptures. He attempted to soften the assertion by arguing that one can locate truth in nature as equally as locating truth in the Scriptures. Ultimately, for Ross, "the Bible teaches a dual, reliably, consistent revelation. God has revealed Himself through the words of the Bible and the facts of nature."[141]

Ross also considered nature to be the sixty-seventh book of the Bible by appealing to twenty-seven passages in the Bible.[142] For example, in Psalm 19:1–4, according to Ross, God speaks through creation.[143] He included Job 12:7 and declared that birds and fish teach about God's creation. In addition, Ross included Psalm 85:11 insisting that truth springs from creation. For Ross, the words of the Bible are God-breathed as stated in 2 Timothy 3:16. The words that God spoke through the work of His hands are also God-breathed. Ross implied that nature could accurately communicate God's mind by means of past, present, and future scientific observations.

Fazale Rana opined that even though creation is a transcendent miracle (God acting outside of matter, energy, space, and time), "the creation event is a testable idea that can fall within the domain of science."[144] Ross added that God creating in six twenty-four-hour days make observational science impossible.[145] PDAC claims that they can know what God intended to communicate through the creation event by reading the Bible and "reading" nature. Ross summarized his belief of two equally authoritative sources of revelation: "anticipat[ing] God's 'two books' will prove consistent internally, externally, and mutually. One provides more detail on the redemptive story,

---

141 Ross, *Creation and Time*, 56.

142 Ross, *Creation and Time*, 56.

143 Ross, *Creation and Time*, 57.

144 Fazale Rana and Hugh Ross, *Origins of Life: Biblical and Evolutionary Models Face Off* (Colorado Springs: NavPress, 2004), 36, 208.

145 Ross, *Creation and Time*, 48.

the other more detail on the creation story, but they speak in perfect harmony. Neither negates or undermines the other."[146]

Definition of יוֹם

Ross asserted that יוֹם (the Hebrew word for *day*) does not mean a twenty-four-hour day in Genesis 1 insisting that יוֹם has a range of meanings. First, יוֹם can mean the period of light as contrasted with the period of darkness. Second, יוֹם can refer to a general, non-descriptive time. Third, יוֹם can signify a point in time. Fourth, יוֹם can indicate a year in the plural. Fifth, יוֹם can mean a twenty-four-hour period of time, which he believed Genesis 1 does not assert.[147] Instead, Ross insisted that יוֹם is frequently interpreted as a long period of time.[148]

Ross added that even when the cardinal (one, two, three, etc.) or ordinal (first, second, three, etc.) numbers are attached to יוֹם such as in Genesis 1:3 (one day), 1:8 (second day), 1:13 (third day), there is "no grammatical rule [that] requires a numbered יוֹם, especially in reference to divine activity, be a twenty-four hour period of time."[149] He provided the example of Hosea 6:2: "He will revive us after two days; He will raise us up on the third day, That we may live before Him." Ross emphasized that Bible commentators "have noted that the 'days' in this passage (where the ordinal is used) refer to a year, years, thousands of years, or maybe more."[150] Ross also maintained, "If Moses wanted to communicate a creation story consisting of six eons, he would have no other option but to use the word יוֹם to describe those eras."[151]

---

[146] Hugh Ross, "Old-Earth (Progressive) Creationism," in *Four Views on Creation, Evolution, and Intelligent*, ed. J. B. Stump (Grand Rapids: Zondervan, 2017), 71.

[147] Hugh Ross, *Genesis One: A Scientific Perspective*, 4th ed. (Covina, CA: Reasons to Believe, 2006), 25.

[148] Ross, *Creation and Time*, 46; Ross, *Genesis One*, 25.

[149] Ross, *Four Views*, 61.

[150] Ross, *Creation and Time*, 47.

[151] Hugh Ross, *Navigating Genesis: A Scientist's Journey through Genesis 1–11* (Covina, CA: RTB, 2014), 35.

Ross rejected the idea that the Hebrew words עֶרֶב (evening) and בֹּקֶר (morning) when added to יוֹם means that a twenty-four-hour cycle had elapsed. The Hebrew word עֶרֶב can mean "sunset" and "end of the day," and בֹּקֶר can mean "sunrise" as well as the "beginning of the day."[152] For this reason Ross argued that "Genesis 1 may well refer to the ending of one time period and the beginning of another, regardless of the length of that period."[153] For example, Ross opined that the phrase "'in my grandfather's day' refers to my grandfather's lifetime; thus, the morning and evening of his day would be his youth and old age."[154] Ross had already concluded that יוֹם could only mean a long period time. He was not interested in the contextual markers to determine meaning of a word. For Ross, the addition of cardinal and ordinal adjectives and the nouns עֶרֶב and בֹּקֶר have limited bearing upon understanding the definition of יוֹם. Hence, he summarized his position like this: when the author of Genesis connects יוֹם to עֶרֶב and בֹּקֶר and a cardinal or ordinal adjective, יוֹם does not mean twenty-four hours, particularly in Genesis 1.

Understanding of the Seventh Day

Another argument that Ross made to defend billions of years in Genesis 1–2 was the belief that the seventh day of the creation event has not ended. He noted the different endings of each day of creation. For days one through six the verses end with the phrase there was evening and morning, while for day seven, the verse ends by stating that God rested on the seventh day from all His work which He had done. Ross argued that the different ending for day seven "strongly suggest that this [seventh] day has (or had) not yet ended."[155]

---

[152] Ross, *Four Views*, 82.
[153] Ross, *Four Views*, 82.
[154] Ross, *Creation and Time*, 46.
[155] Ross, *Creation and Time*, 48–49.

To further his position on the seventh day, Ross added that Psalm 95 and Hebrews 4 affirm that God's seventh day of rest was ongoing: Ross wrote, "The seventh day of the creation week carries on through centuries, from Adam and Eve, through Israel's development as a nation, through the time of Christ's earthly ministry, through the early days of the church, and on into future years."[156] He concluded from these passages that, at minimum, several thousands of years have passed, but most likely billions of years have elapsed. Ross added, "Given the strong parallel structure of the passage, if the seventh day represents a lengthy time period, it seems reasonable that the other days could be lengthy periods as well."[157] Ross had already concluded the days of creation were not six twenty-four hours, thus He continued to look for verses that supported his PDAC view. He presupposition that the earth was 13.2 billion years old led him to find verses he thought supported his position. Ross eliminated any ambivalence by declaring, "an integrative analysis of all these passages leads to the conclusion that יוֹם refers to a long, but finite, time period. This understanding of 'day' yields a consistent reading of all the Bible's creation texts."[158]

Death Before Adam's Sin

Ross believed that death, disease, and bloodshed have always been part of God's creation. He rejected the interpretation that Romans 5:12 affirms that death entered the world because of Adam's disobedience. Ross added, "Paul [in Romans 5:12] clarifies that Adam's sin inaugurated death among humans. Neither here nor anywhere else in the Scriptures does God's word say that Adam's offense brought death to all *life* (emphasis Ross)."[159]

From Ross's perspective, death has existed from the beginning of time. Plants died when the first animals ingested them, and animals have

---

[156] Ross, *Creation and Time*, 49.

[157] Ross, *Genesis One*, 27; Ross, *Four Views*, 80; Ross, *Creation and Time*, 49.

[158] Ross, *Navigating Genesis*, 89.

[159] Ross, *Four Views*, 82.

experienced death for billions of years. He added, "Romans 5:12 addresses neither physical death or soulish death. It addresses spiritual death . . . [Adam] died spiritually [when] he broke his harmonious fellowship with God and introduced the inclination to place one's own way above God's."[160] Death has always existed since God created the heavens and the earth. Ross wrote since "[he] nurtured the seeds of Earth's first life, perhaps re-creating these seeds each time they were destroyed."[161] Although it was hostile during the early events of the earth, God ensured that life would persist, albeit at times by divine intervention (a miracle). Ross based that belief upon the second law of thermodynamics, which states that heat flows from hot bodies to cold bodies:

> A consequence of this direction of heat flow is that, as time proceeds, the universe becomes progressively more mixed or disordered. This increasing disorder, with time, is the principle of decay, also termed "entropy."[162] The law of decay makes possible photosynthesis and all the food photosynthesis provides. It allows us to digest our food. It allowed Adam and Eve, before and after the fall, to perform work. The law of decay brings many more good things, but it also produces inevitable pain, suffering, and death.[163]

For Ross, the decay of plants or the digestion of fruit is not part of the curse; decay began when God implemented photosynthesis or Adam ate a piece of fruit.

For Ross, the bondage that creation has endured and that Paul addressed in Romans 8:20–22 is not the result of Adam's sin. According to Ross, death, disease, and bloodshed is the natural order that God created, because

[160] Ross, *Creation and Time*, 61.
[161] Rana and Ross, *Origins of Life*, 43.
[162] Ross, *Creation and Time*, 66.
[163] Ross, *Navigating Genesis*, 92.

"without decay, work (at least the universe God designed) would be impossible. Without work, physical life also would be impossible, for work is essential to breathing, circulating blood, contracting muscles, digesting food–virtually all life-sustaining processes."[164]

Contrary to Ross, the death that Paul speaks of is the spiritual and physical death that humans experience because of Adam's sin: "Genesis 3 teaches that God judged the nonhuman creation because of Adam's sin."[165] According to Ross, since life began on the third day of creation and Adam was working on the sixth day, Adam's sin could not have inaugurated the decay of plants, for example, which is a form of death. Ross insisted, "process of [death] has been in effect since the universe was created."[166]

Interpretation of Noah's Flood

Ross did not see the flood event to be universal in nature. He stressed, "Worldwide with respect to people and the animals associated with them, which is not to say global."[167] For Ross worldwide meant that which Noah could see, not that the entire earth was covered with water. According to Ross, the flood only covered "the settlements in Mesopotamia and the Persian Gulf Oasis."[168]

Ross provided biblical support for his concept of Noah's flood. First, he argued that in Genesis 41:57 that *worldwide* did not mean every country came to Egypt to purchase food during Joseph's time. He added that all of the peoples lacking food did not include the Australian aborigines and the Inuits, who live in what is now Alaska.

---

[164] Ross, *Creation and Time*, 65–66. Ross never explained why God's very good creation included death, disease, and decay.

[165] Ken Ham, "Young-Earth Creationism," in *Four Views on Creation, Evolution, and Intelligent Design*, ed. J. B. Stump (Grand Rapids: Zondervan, 2017), 25.

[166] Ross, *Creation and Time*, 67.

[167] Ross, *Four Views*, 85.

[168] Ross, *Navigating Genesis*, 149.

Second, in 1 Kings 4:34, the foreign dignitaries came from all the nations to see Solomon. Ross revealed that in 1 Kings 4:31 and 2 Chronicles 9 all the nations extended to Sheba and Arabia rather than every part of the earth. Third, in Luke 2:1, that all of the world that was supposed to register during the reign of Caesar Augustus did not refer, for example, to the Americas. Ross noted that the known world was the Roman Empire rather than every geographical region. Fourth, in Acts 2:5, during the time of Pentecost, every nation under heaven was present. Ross countered that not every people group was present like the Australians or Bolivians.[169]

Ross's point was that the Scriptures allow the word *worldwide* to mean what humans in Israel knew to exist, not necessarily every location on the planet. He stated, "Each of these biblical references to a worldwide occurrence points to an area less than earth's entire surface or entire land area. Therefore, phrases such as 'the entire heavens' and 'the face of the earth' in the context of Noah's Flood may also refer to an area or region smaller than the whole of earth's surface."[170]

Geologically, Ross claimed that there is a lack of direct geological evidence for the flood.[171] The reason is that a flood of limited size could not account for "all of earth's major geological features, [as this] flatly contradicts the physical evidence."[172] Hence, Ross dated the flood during Noah's lifetime as approximately forty thousand years ago.[173]

Ross rejected the existence of sedimentary layers and marine fossils on all seven continents and on the top of multiple mountain ranges, such as the Himalayas, as geological evidence that Noah's flood occurred.[174] According to Ross, the better explanation for finding marine fossils on the Himalayas is due to plate tectonics (the collision of continents). Mountain ranges have

---

169 Ross, *Navigating Genesis*, 146.
170 Ross, *Navigating Genesis*, 147.
171 Ross, *Navigating Genesis*, 156.
172 Ross, *Navigating Genesis*, 155.
173 Ross, *Navigating Genesis*, 156–57.
174 Ross, *Navigating Genesis*, 157.

been rising at a specific rate, and based upon that, are approximately 15 million years old.[175] The sedimentary layers on other continents are from local deluges. Some of those sedimentary layers formed before Noah's flood about 200 million years ago.[176]

In summary, Ross did not believe Noah's flood was global. Instead, it was worldwide to the degree that Noah observed it. For Ross, the scientific evidence supports an old earth. He explained, "a compelling case can be made for seeing the event as worldwide, or universal, with respect to humanity, but less than global with respect to geography.[177]

Context of the Gospel

For Ross, death, disease, and bloodshed was part of God's desired plan.[178] Accordingly, death and decay are not always evil.[179] God declared creation "very good" in the sense that the best possible world exists. Ross declared, "in which God efficiently, rapidly, and permanently conquers evil and suffering while allowing free-will humans to participate in his redemptive process and plan."[180] Adam's sin brought death to humanity but not to the animal kingdom. Instead, animals have always experienced death, disease, and bloodshed.

According to Ross, Christ entered to restore humanity, not necessarily to restore animals and plants. For Ross, the death of nonhuman life for billions of years "blessed humanity with a treasure chest of more than seventy-six quadrillion tons of biodeposits from which to build a global civilization and facilitate the fulfillment of the Great Commission in mere thousands, rather than millions, of years."[181]

---

[175] Ross, *Navigating Genesis*, 159.
[176] Ross, *Navigating Genesis*, 159.
[177] Ross, *Navigating Genesis*, 154.
[178] Ross, *Navigating Genesis*, 75–76.
[179] Ross, *Navigating Genesis*, 75–76.
[180] Ross, *Four Views*, 87.
[181] Ross, *Four Views*, 86–87.

Although tragic, the minimization of human suffering necessarily required billions of years of death and disease of plant, animal, and hominid life. In accordance with Ross, Christ entered a world of death and bloodshed that has always been a part of His creation. Only through the process of death could the gospel be facilitated to reach the maximum number of people to enter into the new heavens and earth in order to be free of disease, bloodshed, and death.[182]

For Ross, the world in which humans are currently residing with pain and suffering "aligns with the day-age vision of the future creation, new heavens and new earth."[183] One day evil and suffering will permanently end. Ross noted, "Death, decay, pain, and darkness (physical and spiritual) will be forever banished."[184]

According to Ross, the eternal state will come, and all forms of suffering will end. In the meantime, humanity waits until the "full number of humans comprising God's kingdom has been granted and received citizenship, evil and suffering will be permanently removed, the universe will have fulfilled its purpose, and we'll be ushered by God into an entirely new and different realm."[185] Ross never explained the origin of the gratuitous suffering for nonhuman life.

Summary of Ross's Day-Age Creationism

Ross insisted that there are sound reasons and reliable evidence that the universe is billions of years old. First, two inerrant sources of revelation apply: the Bible and nature. Both are reliable and will not contradict each other. Second, the Hebrew word יוֹם (translated as *day*) can mean a definite, long period of time, and the nouns עֶרֶב (translated *sunset*) and בֹּקֶר (translated as *sunrise*) have a limited bearing upon understanding the definition of יוֹם in

---

[182] Ross, *Four Views*, 87.

[183] Ross, *Four Views*, 87.

[184] Ross, *Four Views*, 87.

[185] Ross, *Four Views*, 87.

Genesis 1. As a result, Ross concluded that the term *day* in Genesis 1 does not mean twenty-four hours. Third, the seventh day in Genesis does not end with the same "evening and morning phrase" as day one through day six do; thus, there is the possibility that the prolonged nature of day seven could apply to days one through six. Fourth, Noah's flood could not have happened roughly four thousand years ago. The geological evidence of sedimentary layers and marine fossils on top of mountains is consistent with millions of years. Fifth, the second law of thermodynamics requires plants to die and decay, which would mean Romans 5:12 only addresses the spiritual death of humans. According to Ross, not all death and decay are evil.

Ross's arguments are based upon the presupposition that the earth is billions of years old. He derives his presupposition upon the prevailing view that observational science proved the earth is billions of years old; hence, for Ross, the Genesis creation account can only support the PDAC view.

THE CENTRAL TENETS OF SIX-DAY CREATIONISM WITH BRIEF CRITICAL COMMENTARY OF ROSS

Young-earth creationists reject any old-earth theory's conclusions that seek to set the upper limits of the age of the universe and earth much beyond ten thousand years old.[186] They also reject any interpretation of Genesis that endorses a Darwinian evolutionary model allowing for billions of years of a death, disease, and bloodshed cycle before Genesis 1. Young-earth creationists embrace the Six-Day Creation Theory (SDC) as the only view that can accurately describe the Genesis creation account. The theory asserts that God created the universe and the earth throughout six twenty-four-hour periods. Based on other textual markers in Genesis, the universe is thousands of years old.

---

[186] David McGee, "Creation Date of Adam from the Perspective of Young-Earth Creationism," *Answers Research Journal* 5 (2012): 217–30.

Numerous individuals have shaped SDC. For example, some people of influence within the SDC movement are Ken Ham, the founder of Answers in Genesis, Henry Morris III with the Institute for Creation Research, and Carl Wieland with Creation Ministries International. From the leadership of Answers in Genesis, Institute for Creation Research, and Creation Ministries International, other smaller groups organized. John Whitcomb and Henry Morris, who are considered the modern creation movement's fathers, have influenced the three ministries.[187] The SDC affirms that the traditional under-standing of the Genesis creation account is that the Scriptures are unambiguous that the days of the creation week in Genesis 1:1–2:3 are literal, twenty-four-hour days, just like the present duration of the days of the week.

One Primary Source of Revelation

The SDC affirms two sources of revelation, nature and the Bible, but the Bible is the primary source of God's revelation. In other words, theologians should defer to the Bible principally. Nature is a secondary source because it does not compose propositional statements that theologians can evaluate as true or false. Unlike Ross, they would argue that it is simply inaccurate to classify nature as the sixty-seventh book of the Bible: "God's creation speaks to us non-verbally" while "the Scriptures speaks to us verbally and truthfully about so much more . . . creation is cursed, whereas the Scriptures (the written Word) is not."[188]

Jonathan Sarfati added that of the two sources, only the Bible could reveal propositional revelation. Similarly, nature "must be formulated from the observations by *interpreting* them in a framework or *paradigm* (emphasis Sarfati)."[189] Nature does not communicate, "this is what I am saying, or this

---

[187] Terry Mortenson, and Thane H. Ury, ed., *Coming to Grips with Genesis: Biblical Authority and the Age of the Earth* (Green Forest, AR: Master Books, 2008), 8.

[188] Ham, *Four Views*, 19.

[189] Jonathan Sarfati, *Refuting Compromise: A Biblical and Scientific Refutation of Progressive Creationism (Billions of Years) as Popularized by Astronomer Hugh Ross* (Green Forest, Arkansas: Master Books, 2004), 41.

is what I mean after observing me." Instead, all people interpret nature based upon their worldview.

An atheist interprets nature without God, while a theist interprets nature with God. No one is neutral when it comes to comprehending the observations of nature. As a result, there must be an arbitrator who can determine which viewpoint is correct. Only the Bible as a revelation from God can fulfill the role of arbitrator.

Presuppositions, likewise, can influence the interpretations of the Bible (or any text). The difference is that theologians can interpret the Bible (or any text) correctly based upon the laws of logic (which originate from God). For example, the law of noncontradiction affirms that A cannot be A and non-A in the same sense and at the same time.[190]

People assume that communication occurs through writing, as long as the intent of the author is expressed rather than the opposite of what he intended to communicate. The authors of the Bible (divine and human) crafted their thoughts through the means of writing in such a way that the readers could understand them. On the one hand, nature does not express itself through writing; thus, nature's observers must interpret their observations through their presuppositions and then present an interpretation. Since the observer cannot know the interaction of things like God does, the observer hopes his observation is accurate.[191] Since scientific observations continuously change, confidence in interpreting nature is low.

On the other hand, interpreting the Scriptures correctly begins with the claim that the Bible is without error (in the original autographs).[192] God has already revealed what He wants the reader to know. Interpreting the Scriptures is interpreting revelation from God.[193] The reader has confidence

---

[190] Jason Lisle, *Introduction to Logic* (Green Forest, AK: Master Books, 2018), 27.

[191] Jay Adams, *Is All Truth God's Truth?* (Stanley, NC: Timeless Texts, 2003), 4.

[192] R. Albert Mohler, "Confessional Evangelicalism," in *Four Views on the Spectrum of Evangelicalism*, eds. David Naselli and Collin Hansen (Grand Rapids: Zondervan, 2011), 79.

[193] Adams, *Is All Truth God's Truth,* 9.

that he can ascertain God's revelation because the reader is reading the very revelation of God.

To state it another way, nature does not reveal specific revelation while the Scriptures do. Nature reveals limited information, but the Scriptures reveal that which is most important for humankind to know. The observer's interpretative grid mediates nature, but written language mediates the word of God.

In addition, the Bible has no corruption (and the same applies to the ancient copies to the degree that they align with the original). In contrast, nature is corrupted due to the effects of the curse described in Genesis 3. SDC's proponents start with the supremacy of the revelation of the Bible and then look to nature to see how it can complement the Scriptures. Moreover, for SDC, built into this framework of the supremacy of the Bible is the recognition that humanity cannot know everything, particularly the origins of the universe. Ham remarked, "if we start with the someone [God] who knows everything, who does not lie, and who has revealed to us what we need to know,"[194] then the reader can know what happened at the beginning of time when humanity was not present. The SDC view places higher confidence upon accurately interpreting the meaning of the Bible than accurately interpreting the meaning of nature's scientific discoveries. Because the human mind is corrupted and needs divine assistance, the SDC view does not view nature as the Bible's sixty-seventh book.

Definition of יוֹם

According to SDC, based upon the Hebrew language, יוֹם (day) can have five meanings: 1) the period of light (as contrasted with the period of darkness), 2) the period of twenty-four hours, 3) a general, vague *time*, 4) a

---

[194] Ken Ham, *Six Days: The Age of the Earth and the Decline of the Church* (Green Forest, Arkansas: Master Books, 2013), 50.

point of time, and 5) a year.[195] SDC argues that the correct meaning of יוֹם requires an understanding of the context. The context of creation for SDC is the first chapter of Genesis.

SDC asserts, 1) יוֹם always refers to a normal literal day when used as a singular noun; 2) in Genesis 1:1–2:3, יוֹם is used thirteen times in the singular and one time in the plural; 3) when יוֹם is used with עֶרֶב (evening or sunset) and בֹּקֶר (morning or sunrise) it means a literal day; 4) when יוֹם is qualified with a cardinal and ordinal number, the meaning is a literal day; 5) when עֶרֶב and בֹּקֶר are used without יוֹם (thirty-eight times), the meaning of יוֹם is still a literal day; and 6) עֶרֶב and בֹּקֶר are used together with יוֹם six times within Genesis 1:1–2:3 and nineteen times outside of Genesis 1:1–2:3.[196] All these points indicate that the author of Genesis intended to communicate clearly that each day of Genesis 1:1–2:3 was a literal day. SDC emphasizes that Moses, the author-compiler of the Bible's first book, tried to communicate a particular understanding of יוֹם in Genesis 1:1–2:3. He used temporal markers such as "one," and "second," with יוֹם and bounded contextually יוֹם to the words "evening" and "morning." Moses used those words to communicate that each creation day was a literal day.[197]

SDC concludes that assigning any meaning to יוֹם other than a literal twenty-four-hour period of time in Genesis 1 is impossible contextually. Advocates of this view declare that no other exegetical response fits the plain meaning of יוֹם. Had Moses intended to communicate that God created the

---

[195] Leonard Coppes, "852, יוֹם", in *Theological Wordbook of the Old Testament*", eds., R. Laird Harris, Gleason L. Archer Jr., and Bruce Waltke (Chicago: Moody, 1999), 370; Ludwig Koehler, Walter Baumgartner, and Johann Jakob Stamm, *The Hebrew and Aramaic Lexicon of the Old Testament* (Leiden, Netherlands: E. J. Brill, 1994–2000), 399; Tim Chaffey, and Jason Lisle, *Old Earth Creationism on Trial: The Verdict is In* (Green Forest, AR: Master Books, 2008), 25.

[196] Robert V. McCabe, "A Critique of the Framework Interpretation of the Creation Week," in *Coming to Grips with Genesis: Biblical Authority and the Age of the Earth*, eds., Terry Mortenson and Thane H. Ury (Green Forest, AR: Master Books, 2008), 225–28.

[197] Ham, *Four Views*, 21.

earth in six twenty-four-hour periods, what words or phrases would he choose to use? SDC unabashedly answers that Moses's exact choice of words in Genesis 1:1–2:3 support the view that יוֹם means a literal day lasting approximately twenty-four hours.

Understanding the Seventh Day

According to SDC, the seventh day of creation has ended, thus, it has not continued for the last six thousand years as PDAC purports.[198] SDC provides several arguments for a defense. First, "the text [of Hebrews 4:3–5] does not say that the seventh day of the creation week is continuing to the present day. It merely reveals that God entered His rest on the seventh day."[199] The author of Hebrews does not state in this section that somehow God's Sabbath rest has continued until the present; rather, he links "God's Sabbath-rest at the time of Creation with the rest that Israelites missed in the desert."[200] There is a future rest that the original audience could miss, but that rest is not a continuation of the seventh-day rest.

Second, SDC affirms that the seventh day must be literal "because Adam and Eve lived through it before God drove them out of the garden. Surely, He would not have cursed the earth during the seventh day, which He blessed and sanctified."[201]

Third, SDC rejects the claim of PDAC that Hebrews 4:3–5 affirms that the seventh day must be a long time because the phrase "evening and morning" is not included. Instead, SDC argues that if the exclusion of the phrase "allows the seventh day to be longer, then this is really an unintentional admission that the first six days were literal 24-hour days."[202]

---

[198] Norman Geisler, *Systematic Theology*, vol. 2 (Minneapolis, MN: Bethany House, 2003), 643.

[199] Chaffey and Lisle, *Old Earth*, 51.

[200] Zane Hodges, "Hebrews," in *The Bible Knowledge Commentary*, eds., J. Walvoord and R. Zuck, vol. 2 (Wheaton, Illinois: Victor Books, 1985), 788.

[201] John Whitcomb, "The Science of Historical Geology," *Westminster Theological Journal* 36 (1973): 65–77.

[202] Chaffey and Lisle, *Old Earth*, 52.

In other words, by the interpretative method of PDAC, if the omission of the phrase "evening and morning" for the seventh day of the creation week is evidence to suggest that the seventh day can be indefinite, then the inclusion of that phrase "evening and morning," which is bounded to days one through six of the creation week, should be evidence to suggest that those days are definite.

Davidson remarks, "the references to 'evening' and 'morning' together, outside of Genesis 1, invariably, without exception in the Old Testament (fifty-seven times total–nineteen times with יוֹם, or 'day,' and thirty-eight without יוֹם) indicate a literal solar day."[203] At best, according to PDAC, if their interpretation is correct, day seven could be indefinite; however, days one through six are literal twenty-four-hour days. The lack of grammatical evidence undermines Ross's purpose of transferring the so-called "indefiniteness" of day seven to days one through six.

Death after Adam's Sin

Before the end of the sixth day of creation, God had declared multiple times that what He had created was good. Furthermore, God declared that all He had created was very good at the end of the sixth day of creation. James Swanson commented that the word מְאֹד, translated *very*, carries with it the idea of "greatly, utterly, i.e., pertaining to a high point on a scale of extent."[204] God's creation pinnacle was the end of the sixth day. Those who espouse the SDC theory believe the Scriptures communicate no death or disease before Adam and Eve's sin.[205] It would seem odd for God to declare His creation on days one through five good and then highlight day six as very good while death, disease, and bloodshed had been occurring for millions of years.

---

[203] Richard Davidson, "The Genesis Account of Origins," in *The Genesis Creation Account: And Its Reverberations in the Old Testament*, ed., G. A. Klingbeil (Berrien Springs, MI: Andrews University Press, 2015), 78.

[204] James A. Swanson, *Dictionary of Biblical Languages with Semantic Domains: Hebrew* (Oak Harbor, WA: Logos Research Systems, 1997).

[205] Chaffey and Lisle, *Old Earth*, 71.

Genesis 3 asserts that the ground was not cursed until Adam and Eve sinned. Verses seventeen and eighteen affirm that creation was not subject to death, bloodshed, or disease: "Cursed is the ground because of you; in toil you will eat of it all the days of your life. Both thorns and thistles it shall grow for you." Paul's commentary in Romans about the Fall supports the teaching that the curse came after sin (Rom. 8:20). The only place in the Scriptures that designates what could be described as a historical global-scale curse is in Genesis 3. Death, disease, and bloodshed were not part of the original creation event. On the contrary, the earth was *very good* and functioned exactly how God designed it.

Interpretation of Noah's Flood

A majority of proponents of SDC, including geologist Andrew Snelling, affirm Noah's flood occurred approximately seventeen hundred years after the creation week, and it lasted approximately 371 days.[206] Snelling presents seven evidences to affirm the Genesis flood.[207] Ken Ham and Tim Lovett concluded that God used two water sources to flood the earth: "the windows of heaven" and "fountains of the great deep."[208] Moses, the author of Genesis, listed two sources (Gen. 7:11) and commented that the waters above the earth rained for forty days while the water below the earth's crust burst forth for one hundred and fifty days.[209] The flood was not local, instead, it was global and covered all the earth.

---

[206] Andrew Snelling, *Earth's Catastrophic Past: Geology, Creation, and the Flood* (Dallas: Institute for Creation Research, 2009), 279; Ken Ham et al., *A Pocket Guide to the Global Flood* (Hebron, KY: Answers in Genesis, 2009), 94–95; William D. Barrick, "Noah's Flood and its Geological Implications," in *Coming to Grips with Genesis: Biblical Authority and the Age of the Earth*, eds., Terry Mortenson and Thane H. Ury (Green Forest, AR: Master Books, 2008), 272–73.

[207] Andrew Snelling, "Global Evidences of the Genesis Flood," *Answers Magazine*, July 1, 2021, https://answersingenesis.org/the-flood/global/evidences-genesis-flood/.

[208] Ham et al., *A Pocket Guide*, 11.

[209] Ham et al., *A Pocket Guide*, 9–13.

SDC argues that eight times Genesis 7:17–23 describes the waters prevailing upon the earth that covered all the high hills and destroyed all air-breathing, land-dwelling creatures. If the author wanted to convey a local flood, the language in Genesis communicated the opposite. Advocates of SDC refute the local flood theory by revealing the scriptural problems that result from embracing the local flood view.

First, if the flood was local, Noah and his family could have relocated to higher ground. Second, the pairs of each land animal did not need to enter the ark. Third, God promised not to send another flood, yet He has allowed local floods to continue presently. Fourth, if local, then God did not fulfill His promise to destroy all of humanity. Fifth, Jesus believed the flood killed all of humanity (minus the eight).[210]

Summary of Six-Day Creationism

Advocates of SDC insist that there are sound biblical reasons and interpreted evidence that presents a case to conclude the universe is not much older than six to ten thousand years. First, the primary source of God's revelation, the inerrant word of God, is magisterial, but nature, a cursed source, is ministerial. Therefore, believers ought to interpret the scientific evidence in light of Genesis 1. Ham summarized the first reason: "The difference between young-earth creationists and all our Christian and non-Christian opponents is that we accept God's eyewitness testimony in the Scriptures and use it to interpret the physical evidence that we see in the present."[211]

Second, יוֹם always means a twenty-four-hour literal day when combined with עֶרֶב and בֹּקֶר, and qualified with a cardinal and ordinal number. Third, the seventh day of creation was a literal day, lasting approximately twenty-four hours. Fourth, death, disease, and bloodshed did not begin until Adam

---

[210] Ken Ham and Bodie Hodge, *A Flood of Evidence* (Green Forest, AR: Master Books, 2016), 94–95; Ham et al., *A Pocket Guide*, 16.

[211] Ham, *Four Views*, 34.

sinned. Finally, Romans 8:20 teaches that creation waits for God to release it from the curse. The only place in Genesis that describes a curse is Genesis 3 when Adam sinned.

Fifth, most advocates of SDC affirm a timeline of Noah's flood that occurred approximately 1,656 years after creating the world. The flood was global in scope and not local. It killed all air-breathing, land-dwelling animals and humans except those preserved in the ark.[212] Advocates of SDC and advocates of PDAC have divergent explanations regarding the creation event and Noah's flood.

DETAILED CRITICAL COMMENTARY
OF HUGH'S ROSS'S POSITION WITH SDC RESPONSE

Supremacy of Special Revelation or Natural Revelation

Ross argued that "if all of creation were completed in six twenty-four-hour days, the most sophisticated measuring techniques available, or even foreseeably available, would be totally incapable of discerning the sequence of the events. Thus, a major use of the chronology would be thwarted."[213] In other words, Ross is arguing the current (or future) model of scientific observation cannot validate the SDC view, even if correct; therefore, the age of the universe would be unknowable. Theologians should reject the SDC model. Ross's analysis is a false analogy because the SDC view claims that one can know the universe's approximate age based on the textual clues left within Genesis 1–11.[214]

Ross affirmed that scientists' interpretations regarding the universe's age based upon their observations of nature are correct beyond a reasonable doubt. For this reason, theologians must adjust their interpretations of the Scriptures. Ross added, "God's revelation is not limited exclusively to the

---

[212] Ham et al., *A Pocket Guide*, 94–95; Ham, *Four Views*, 28–29.

[213] Ross, *Creation and Time*, 48.

[214] Kurt Wise, *Faith, Form, and Time* (Nashville: B & H, 2002), 50–52.

Bible's words. The facts of nature may be likened to a sixty-seventh book of the Bible."[215] Ross's comments mean nature is an inerrant revelation from God, which is reliable like the Bible. Although this writer agrees with the PDAC view that God teaches believers things through nature, this writer cannot accept the overall conclusion that nature is equivalent to a sixty-seventh book of the Bible. The primary reason is that nature is cursed as described in Genesis 3.

As the apostle Paul described in Romans 8:20–22, God subjected nature to a form of emptiness. It was enslaved and waits for God to set it free. Although nature depicted as enslaved is a literary device of personification, Paul pointed to a fundamental change in nature after God pronounced judgment upon Adam, Eve, the serpent, and the whole earth. Thus, since God has cursed nature, nature can distort the theologian's observations of nature when interpreting it.

Second, the curse has affected humanity's mind as described in Genesis 3. As the apostle Paul describes in Romans 1:18–32, the human mind suppresses the truth, is foolish, and promotes atheism.[216] The result of God giving humanity what it wants is that it worships creatures rather than Him, embraces sexual behavior contrary to biology, and revels in every form of wicked behavior possible. Consequently, since the Genesis 3 curse has affected nature and the human mind, the combination of a defective nature and a defective human mind would, at times, produce faulty observations and faulty interpretations.[217] For this reason, the human mind, which interprets scientific observations, can be wrong.

Historically, believers have struggled to interpret the Scriptures accurately. Even when they encounter a difficult passage to interpret, however, they have always had the confidence that situated in the text is the answer. The

---

[215] Ross, *Creation and Time*, 56.

[216] R. C. Sproul, *Defending Your Faith: Introduction to Apologetics* (Wheaton, IL: Crossway, 2003), 159–63.

[217] Terry Mortenson, ed., *Searching for Adam: Genesis and the Truth About Man's Origin* (Green Forest, AR: Master Books), 461.

theologians and scientists of the SDC view recognize that they can be in error at times, but they can always return to the Scriptures to test their views. For advocates of SDC, Genesis is the only location for the correct interpretation of the creation account.

The PDAC view has a more difficult task. Advocates of PDAC observe nature and interpret with their minds (which have been affected by the Genesis 3 curse) to draw their conclusions. They purport in theory to give supremacy to the Scriptures, but in practice, nature is equal to the Scriptures and, at times, superior to the Scriptures.

Nature is not perfect like the Scriptures thus not the standard. Consequently, nature is incapable of being the sixty-seventh book of the Bible, and instead, it is subject to the Bible. If Ross were consistent, he would need to submit to the Scriptures (the sixty-six books of the Bible) when in conflict with scientists who make interpretations of their observations of nature.[218]

To further understand Ross's view of special and general revelation, one needs to understand his view of miracles. According to Ross, there are two kinds of miracles in the Bible: testable and non-testable.[219] Testable miracles are the events of Genesis 1–11, but non-testable miracles are cases like the virgin birth, resurrections, and turning water into wine.

Ross also placed miracles into three other categories: transcendent, transformational, and sustaining.[220] According to Ross, transcendent miracles are God's acts creating space-time and physical laws, which Genesis 1:1 primarily describes. Transformational miracles are God's acts of working with preexisting materials to fashion life on earth and breathe life into humanity, which Genesis 1:2–2:3 describes. Sustaining miracles are God's

---

[218] Ross, *Creation and Time*, 58. Ross asserted that no contradiction exists between the record of nature and the Bible.

[219] Ross, *Navigating Genesis*, 15.

[220] Ross, *Four Views*, 74.

acts to ensure life continues through harsh conditions for millions and billions of years.[221]

The difficulty with the last category of miracles, as Ross presented it, is that the Genesis account nowhere indicates harsh conditions before Adam's sin. On the contrary, Genesis describes everything that God completes on each day as good. It would seem that Ross has borrowed his creation account from Darwin's writings instead of Moses's writings. [222]

Based upon these various labels, non-testable and transcendent miracles would seem to be identical classes; testable and transformational miracles are another class, and sustaining miracles would be in a class by themselves. Sustaining miracles would appear to be closer to the theological concept of the providence of God. If God did not sustain the earth for billions of years, then this class of miracles would be nonexistent. According to Ross, all the miracles in Genesis 1, minus Genesis 1:1, are testable and are within the bounds of scientific inquiry.

This researcher has two concerns. First, Ross believed he could determine the age of the universe after the creation event, but he did not have another universe with which to compare. Accordingly, Ross has no starting point to determine the age of the universe. For example, one can have a high degree of confidence that another person looks like he is in his forties. Certain physical features may confirm this belief (e.g., some gray hair, wrinkles around the eyes, and a general elasticity of the skin). In a similar manner, Ross assumed certain features in the universe as indicators of age. However, what Ross has overlooked are the initial conditions of the universe immediately after God created it. For example, when God created Adam, he may have looked like he was a thirty-year-old adult, but really he was only a day old. In other words, God created Adam as a mature adult.[223]

---

[221] Ross, *Four Views*, 74.
[222] Darwin, *The Origin of Species*, 60.
[223] Chafey and Lisle, *Old Earth Creationism*, 150.

From the perspective of modern observers, the initial conditions of Adam would lead them to think Adam had existed for about thirty years. The observation would be wrong. Chaffey and Lisle add, "age is an indication of history–not appearance."[224] Adam had no history, thus, he "looked" thirty; he was actually one day old.

Since Ross embraced the presupposition that he could determine age based upon previous observations, his conclusions are wrong. He had no way of knowing the initial conditions of the universe; thus, Ross had no way of determining the age of the universe. What he needs is a written record of the time when he was not present to inform him of how much time had elapsed. This record exists in the form of the Bible, and specifically Genesis 1–11.

Secondly, Ross redefined the term *miracle* with his invented categories (testable, transcendent, sustainable, etc.) and created an unrecognizable definition. He divided the term miracle into various categories that suited his purpose in justifying his belief that he could date the age of the earth. Richard Purtill commented that a consensus understanding of a *miracle* is "an event in which God temporarily makes an exception to the natural order of things, to show that God is acting."[225] Miracles are extraordinary, unlikely, and irregular.[226] Geisler added,

> [I]t is not enough to define a miracle as an exception to the general pattern of events. This characteristic merely indicates that the event is a non-natural one; [and] there are other possibilities within the category of non-natural or unusual events: anomalies, magic, alien beings, demonic activity, and even providential activity. The

---

[224] Chafey and Lisle, *Old Earth Creationism*, 149.

[225] Richard L. Purtill, "Defining Miracles," in *Defense of Miracles: A Comprehensive Case for God's Action in History*, eds., R. Douglas Geivett and Gary R. Habermas (Downers Grove, IL: InterVarsity, 1997), 62–63; William Lane Craig, *Reasonable Faith: Christian Truth and Apologetics* (Wheaton: Crossway Books, 2008), 253.

[226] John M. Frame, *Apologetics: A Justification of Christian Belief* (Phillipsburg, New Jersey: P & R, 2015), 145–47.

characteristics of a true miracle are unusualness, immediateness, purposefulness, and moral goodness.[227]

Geisler narrowed the concept of a miracle only to include the unique, unusual, and purposeful action by God.

Ross modified the understanding of a miracle to exclude the creative account of Genesis 1:2–2:4. He took the pericope of Genesis 1:2–2:4, in which God declared that He had created supernaturally (by means of a miracle), and redefined the supernatural creative event in order to create a category that conveniently affirmed his position. Ross's hermeneutic is the literal, historical, grammatical, and canonical *supervised-by-scientific observations* method.[228] In other words, he affirmed a literal interpretation but ignored the passage's context. Based upon his presupposition that the scientific community has settled the universe's age to billions of years, Ross formed his view.[229] In the end, Ross becomes the final arbitrator of the origin debate.

Divergent Hermeneutics

Ross interpreted Genesis 1–11 with a different hermeneutic than SDC, thus, his interpretations were vastly different than the interpretations of SDC. Ross listed the possible literal meanings for the word יוֹם when attached with the adjectives *first, second, third,* and with the nouns *evening* and *morning.* Then he found what he thought was an exception to that literal meaning in Hosea's book.[230] The prophet Hosea wrote, "He will revive us after two days; He will raise us up on the third day, that we may live before Him (Hos. 6:2)." Ross commented, "for centuries Bible expositors have noted that 'days' referred to in this passage (where both cardinal and ordinal numbers are connected

---

[227] Norman L. Geisler, *Christian Apologetics* (Grand Rapids: Baker, 2013), 319.

[228] David McGee, "Critical Analysis of Hugh Ross' Progressive Day-Age Creationism Through the Framework of Young-Earth Creationism," *Answers Research Journal* 12 (2019): 64.

[229] Ross, *Genesis One*, 25. Ross, *Four Views*, 73.

[230] Ross, *Creation and Time*, 46.

with םוֹי) represents years, perhaps as many as a thousand or more."[231] At first glance, Ross made an argument of an interpretation of םוֹי with both a cardinal and ordinal number that could mean longer than twenty-four hours. Understanding the context of Hosea 6:2 demonstrates that Ross did not find a text that proved his assumption.

Contextually, Hosea's book focuses on the coming judgment of the northern tribes of the nation of Israel by the hands of the Assyrians in 722 BC.[232] Israel was guilty of breaking Yahweh's laws by worshipping false gods in the form of idols and displaying injustice to the poor. Yahweh commanded Hosea the prophet to marry a harlot. The harlot represented unfaithful Israel, but Hosea represented Yahweh. As unfaithful as Hosea's harlot wife was to him, so Israel was unfaithful to Yahweh. Toward the end of Hosea 4:1–6:3, after Hosea charged that Israel was guilty of prostituting itself with the surrounding nations by worshipping their gods instead of Yahweh, Hosea prophesied that the nation of Israel would return to Him in repentance. After repenting, Yahweh would heal them.

Within the context of Hosea 6:2–3, Yahweh promised to restore the Israelites quickly within two or no more than three days. Commentators have noticed the literal interpretation of םוֹי with the cardinal and ordinal number. For instance, John Lange remarked, "two and three days are very short periods of time; and the linking of two numbers following the one upon the other, expresses the certainty of what is to take place within the period named."[233] In addition, Hans Wolff affirmed, "the ancient song in vv. 1–3 [of Hosea 6] merely voices the expectation that a sickly nation will be put on the road to recovery by Yahweh, and in the shortest possible time. The set

[231] Ross, *Four Views*, 81.

[232] Ed Hindson and Gary Yates, *The Essence of the Old Testament: A Survey* (Nashville: B & H Academic, 2012), 369–70.

[233] John Peter Lange, *A Commentary on the Holy Scriptures: Hosea* (Bellingham, Washington: Logos Bible Software, (1899, 2008), 61.

length of time, "after two days, on the third day."[234] Robert Chisholm emphasized the future of this prophecy, declaring, "these verses record the words the penitent generation of the future will declare as they seek the LORD," and the "equivalent expressions, after two days and on the third day, refer to a short period of time."[235]

All three commentaries affirm the expression as a literal time period of two or three days (i.e., a brief time). Accordingly, the text does not allow יוֹם in Hosea 6:1–3 to mean thousands, millions, or billions of years as Ross argued. In other words, the textual evidence does not claim that the universe is billions of years old. Chisholm remarked, "The promise only makes sense when we take the days literally and take the phrases as meaning 'quickly.'"[236]

Ross attempted to take the lack of fulfillment of Hosea 6:1–3 (Israel had yet to repent as a nation) and show that the use of יוֹם in this passage with cardinal and ordinal numbers, plus the length of time since Hosea composed this passage (approximately 2,700 years old and counting), justified him to pronounce that all the uses of יוֹם in Genesis 1:1–2:4 could be long periods of time extending into billions of years.

For the sake of argument, even if Ross could establish that this prophetic passage used יוֹם in a non-literal sense (i.e., not twenty-four hours), this passage would not overrule how Genesis 1 uses the term יוֹם. When the Scriptures use יוֹם with a number, particularly with the phrase *evening* and *morning*, יוֹם means a literal day. In other words, יוֹם (like any other word), derives its meaning from the surrounding context, thus, the context of Hosea 6:1–3 defines the meaning of יוֹם. Hence, when one applies the same rules of

---

[234] Hans Walter Wolff, *Hermeneia: Hosea: A Commentary on the Book of the Prophet Hosea,* trans. Gary Stansell (Philadelphia: Fortress, 1974), 118.

[235] Robert Chisholm, Jr., "Hosea," vol. 1, in *The Bible Knowledge Commentary: An Exposition of the Scriptures,* eds., John F. Walvoord and Roy B. Zuck (Wheaton: Victor Books, 1985), 1393.

[236] Ham, *Four Views,* 21.

interpretation to Genesis 1:1–2:4, the context defines יוֹם, which determines the meaning of יוֹם in the creation account.

DISTANT STAR LIGHT PROBLEM

PDAC: Distant Star Light and the Age of the Universe

Ross's philosophical assumptions dictated how he interpreted Genesis 1–11. Generally, for Ross, natural revelation superseded special revelation because the creation event is testable (minus Genesis 1:1).[237] He contended that there is "evidence of a cosmic beginning in the finite past–only 13.8 billion years ago."[238] One primary reason Ross believed the universe's age was approximately 13.8 billion years old was because of distant starlight.[239] The distant starlight problem is one of the most challenging rebuttals for the six-day creationist to answer and one that gives the most robust evidence that the universe is billions of years old. Distant starlight would appear to indicate that PDAC is a more accurate view.

The stars are far away, and their light is too distant to reach earth in 6,000 to 10,000 years as the SDC view claims; therefore, the critics claim the universe must be older than thousands of years, and the SDC view cannot be correct. Described in more detail, the distance from the farthest observed stars to earth is billions of light-years. A light-year is not a unit of time but the distance that light can travel in one year equal to 5.88 trillion miles.[240] The farthest distance of observable stars is calculable, and the rate of the speed of light is constant at approximately 186,000 miles per second. To determine how long it would take, in years, for light to travel from the farthest stars, one could calculate the distance from those stars to earth and divide the distance by one light-year. For example, Alpha Centauri, the next nearest star

---

[237] Rana and Ross, *Origins of Life*, 36.

[238] Ross, *Navigating Genesis*, 15.

[239] Ross, *Creation and Time*, 92–95; Ross, *Navigating Genesis*, 161–64.

[240] Danny Faulkner, "A Proposal for a New Solution to the Light Travel Time Problem," *Answers Research Journal* 6 (2013): 279–84.

system to the Solar System, is approximately 4.3 light-years away from earth (25 trillion miles divided by 5.88 trillion miles).[241]

According to PDAC, light from the most distant stars (i.e., galaxy MACSO647-JD) traveled 13–14 billion light-years to reach earth.[242] The distance requires more time than SDC claims the biblical chronology allows. Ross's assertion required a rebuttal from six-day creationists. Six-day creationists provided a response to Ross's concern, which they believe gave a reasonable solution to how the light from the most distant stars has reached earth in a matter of thousands of years.

SDC: Distant Star Light and the Age of the Universe

First, six-day creationists affirm that the distance from the farthest galaxies is accurate.[243] Second, six-day creationists have advocated several views that can answer the distant starlight problem.[244] Third, the SDC view has continued to critique itself by explaining each proposal's advantages and disadvantages. Of the various solutions proposed by SDC, two of the more popular views that six-day creationists espouse are Humphreys's white hole cosmology view and Lisle's anisotropic synchrony convention view.[245]

In 2008, Humphreys modified his White Hole Cosmology view with a new time dilation model, which he called achronicity, or "timelessness."[246] He amended his previous view because he did not believe the previous

[241] Larry Vardiman and D. Russell Humphreys, "A New Creationist Cosmology: In No Time at All Part 3," *Acts & Facts* 40, no. 2 (2011): 12–14.

[242] Ross, *Creation and Time*, 98.

[243] Jason Lisle, *Taking Back Astronomy: The Heavens Declare Creation and Science Confirms It* (Green Forest, AR: Master Books, 2012), 30.

[244] Faulkner, "A Proposal," 279–84.

[245] Much of the "SDC: Distant Star Light and the Age of the Universe" section, the writer adapted from his article: David McGee, "Critical Analysis of Hugh Ross' Progressive Day-Age Creationism Through the Framework of Young-Earth Creationism," *Answers Research Journal* 12 (2019): 60–63.

[246] D. Russell Humphreys, "New Time Dilation Helps Creation Cosmology," *Journal of Creation* 22, no. 3 (2008): 84.

hypothesis provided a solution to allow enough time dilation for nearby stars and galaxies. Moreover, his metric was too complex to analyze thoroughly.[247]

The modified view's thesis is that at the beginning of the creation event, "the deep" of Genesis 1:2 would have created a dent in space such that conditions near the dent would have caused time and all physical processes to stop. Humphreys suggested that "the deep" would have had a mass "in the order of twenty times that of all galaxies within the viewing rage of the Hubble space telescope [and] would have been in the shape of a ball a few light-years diameter." [248]

Humphreys opined that during the second day of creation, God separated "the deep" with the substance in Hebrew called *raqia*. At the center of "the deep" was a substantial body of spherical water called earth. God separated the remaining waters from the waters on earth by a substance called *raqia*, translated as *firmament* or *the expanse*. The *raqia* spread out spherically. As a result, ice particles surrounded the universe at the end edge of the universe.

Humphreys's analogy was to think of a helium balloon with a marble fixed at the center (or near the center). The rubber material represents "the deep," the marble represents the earth, and the helium represents the *raqia*.[249] In other words, during the second day of creation, God created and expanded the universe with the substance *raqia* (similar to the material scientists call space) and placed earth (a watery spherical mass at this time) at the center or near the center of the universe. As an illustration, Humphreys imagined space representing a trampoline and the universe representing a massive metal ring (the edge was the ice particles of "the waters above," including the *raqia*), creating a spherical indentation. Lying near the center of the metal ring was a pebble representing earth.

---

[247] Larry Vardiman and D. Russell. Humphreys, "A New Creationist Cosmology: In No Time at All Part 1," *Acts & Facts* 39, no. 11 (2010): 12–15.

[248] Vardiman and Humphreys, "A New Creationist–Part 1," 12–15.

[249] Vardiman and Humphreys, "A New Creationist–Part 1," 12–15.

Humphreys suggested that the mass of space, "the deep," spread out with the *raqia*. Its edge (represented by the metal ring) affected time. He declared, "the distribution of mass controls the fabric of space, the fabric of space controls the speed of light, and the speed of light controls time. Time is speeded up or slowed down throughout space according to the distribution of mass."[250]

According to Humphreys, as God was stretching the *raqia*, the powerful mass of "the deep" created the gravitational pull. In effect, this gravitational pull stopped time. While God created the stars and galaxies inside the ring, the light arrived instantaneously on earth. As God stretched the fabric of space, the light trajectory was also stretching. This would account for the redshifting of the light waves.[251]

Humphreys has sought to explain the redshift in light that indicates the stars are billions of light-years away from earth. To explain, redshifting of light, from the perspective of the earth-bound observer, indicates light has traveled long distances. Light has a spectrum of colors. One can see those variations of color in a rainbow. When light moves away from an object, the light is shifted to the spectrum's red end because its wavelengths get longer. As an object moves closer, the light moves to the spectrum's blue end because its wavelengths get shorter.[252]

Humphreys also suggested a second time-dilation event during the Genesis flood. He indicated that if Noah could have seen the night sky (which he could not due to too many clouds from the monumental flood rains), that "[Noah] would have seen the galaxies grow older by about 500 million years."[253] Humphreys advocated two gravitational time dilations that could have occurred, one at creation and the other at the flood, explaining

---

[250] Larry Vardiman and D. Russell Humphreys, "A New Creationist Cosmology: In No Time at All Part 2," *Acts & Facts* 40, no. 1 (2011): 12–14.

[251] Vardiman and Humphreys, "A New Creationist–Part 2," 12–14.

[252] John G. Hartnett, "Speculation of Redshift in a Created Universe," *Answers Research Journal* 8 (2015): 77–83.

[253] Vardiman and Humphreys, "A New Creationist–Part 3," 12–14.

how light traveled from the distant stars to earth. The two gravitational time dilations, he believed, can provide a reasonable response to the distant starlight problem that Ross purported was insurmountable.

In 2001, Lisle proposed a view under the pseudonym Robert Newton. He suggested that there are two conventions of time referred to as *observed time* and *calculated time*. Jason Lisle commented that o*bserved time* is when one sees an event, and *calculated time* "is calculated by subtracting the light travel-time (distance to the event divided by the speed of light) from the observed time."[254]

He asserted that Genesis 1 described the creation of the sun, moon, and stars on day four from *observed time*. Lisle quickly pointed out that *calculated time* would imply that God created the sun, moon, and stars billions of years *before* the light would reach earth on day four. Lisle's implication would contradict the literal interpretation of Genesis 1. His solution was to remind the readers that based upon Einstein's Special Relativity Theory, "the motion of the observer affects the measurement of time."[255] Thus, at *calculated time*, light travels at a constant speed of approximately186,000 miles per second, while at the *observed time*, light travels at various speeds dependent upon the observer's location. Lisle added, "[it] does not appear to be any way to test the unidirectional speed of light empirically."[256] Which *time* is correct? Lisle argued that both are correct. An analogy would be the English and metric systems of measurement. Both are conventions of measurement, and scientists do not emphasize that one convention is correct to the exclusion of the other.[257]

Lisle added that the Bible used *observed time* because Moses could not have known *calculated time*. Moses did not know the speed of light or the distances

---

[254] Robert Newton, "Distant Starlight and Genesis: Conventions of Time Measurement," *Journal of Creation* 15 no. 1 (2001): 80.

[255] Newton, "Distant Starlight and Genesis," 81.

[256] Newton, "Distant Starlight and Genesis," 85.

[257] Jason Lisle, "Anisotropic Synchrony Convention–A Solution to the Distant Starlight Problem," *Answers Research Journal* 3 (2010): 206.

of the farthest stars. Thus, when Moses recorded the creation event, he described the *observed time* of the stars' arrival.

Lisle suggested that if Moses were present on day four, he would have seen the stars instantaneously as God created them. Lisle was aware that this view might seem similar to PDAC. He remarked, "the only similarity—this idea of 'billions' of years—merely comes from the way in which we have chosen to define time, and does not reflect the duration of any actual process."[258] In 2010, Lisle augmented his view by positing that six-day creationists could understand the creation event from two time conventions. Convention one is the time from the perspective of day four on the earth during the creation event. Convention two is the time from the perspective of the distant stars during the creation event.

Lisle also affirmed that the speed of light is constant, based upon a round trip. The round trip is when astrophysicists bounce light off a mirror and return to its source location, where they measure the constant speed of light. What is unique to Lisle's revitalized view is that scientists do not know the one-way speed of light. He argued, "the speed of light in any one direction is not necessarily constant. As counter intuitive as it may seem, the one-way speed of light is not a constant of nature, but is a matter of convention."[259]

According to Lisle, light could have traveled on day four from the most distant stars and instantaneously arrived on earth, as Genesis 1:14–15 appears to indicate. Lisle commented, "it is already well-established that clocks tick slower as they approach the speed of light and would stop completely if they could attain the speed of light. So, from light's point of view (imagine that we could travel alongside the light beam), every trip is instantaneous anyway."[260] If Lisle's view is correct, then the distant starlight problem is no longer an issue to explain because light leaves the newly created stars on day

---

[258] Newton, "Distant Starlight and Genesis," 84.
[259] Lisle, "Anisotropic Synchrony Convention," 199.
[260] Lisle, "Anisotropic Synchrony Convention," 202.

four at the speed of light and arrives essentially instantaneously to earth on the same day.[261]

Summary of Distant Starlight Problem

Humphreys's and Lisle's views provide a possible solution to combat the distant starlight problem that advocates of PDAC present. Humphreys suggested two gravitational time dilations could have occurred. First, he proposed a gravitational time dilation at creation.[262] Secondly, he suggested a gravitational time dilation during the flood. Each event can potentially explain how light traveled from the distant stars to earth. The two gravitational time dilations, he believed, could provide a reasonable response to the distant starlight problem that Ross purported was insurmountable.

Lisle verified that the speed of light based upon a round trip is constant.[263] He also argued that the one-way rate of the speed of light could be instantaneous.[264] Lisle suggested that he proposed a solution for the distant star-light problem because light could have left the newly created stars on day four at the speed of light and arrived essentially immediately to earth on the same day.

Humphreys and Lisle are not the only ones to offer solutions to the distant starlight questions from PDAC's proponents. Danny Faulkner has proposed a "dasha" solution.[265] Not all six-day creationists believe that Humphreys or Lisle offer the best solutions to explain the distant starlight problem. Humphreys and Lisle, however, have provided scientific-based solutions to the distant starlight problem.

---

[261] For detailed mathematical analysis of Lisle's ASC view, Jason Lisle, *The Physics of Einstein: Black holes, time travel, distant starlight, E=mc²* (Aledo, TX: Biblical Science Institute, 2018), 1–282.

[262] Faulkner, "A Proposal," 279.

[263] Lisle, "Anisotropic Synchrony Convention," 202.

[264] Lisle, "Anisotropic Synchrony Convention," 199.

[265] Danny Faulkner, "Solving the Light Travel Time Problem," *Answers in Depth* 16 (2021).

All six-day creationists conclude that there is a solution to explain the distant starlight problem that, when discovered, will be consistent with the Genesis creation account that places an upper limit of the universe at approximately ten thousand years old. In the meantime, the proposals of Humphreys and Lisle provide a reasonable solution for proponents of PDAC to explain the arrival of the most distant starlight.

Radiometric Dating

Ross argued for an old-earth in part because the dating of specific radioactive decay rates provides scientific evidence that the earth is billions of years old.[266] Some of the radioactive elements are potassium to argon, uranium to lead, and rubidium to strontium.[267] Like the distant starlight problem, six-day creationists appear to have a problem with radioactive decay rates because their proposed ages of decay are hundreds of millions of years old. Six-day creationists refute the claims of advocates of PDAC and provide evidence that the earth is thousands, not billions of years old.

Ham and Hodge explained radiometric dating in a digestible way for people who do not have scientific backgrounds. They remarked, "A radiometric dating method requires a radioactive (an element that wants to break down into another element) material A (the parent) into material B (the daughter). For example, a radioactive form of potassium (the parent) wants to break down into argon (the daughter)."[268] Radiometric dating requires an unstable (parent) element such as potassium or rubidium that will break down into a stable (daughter) element such as argon or strontium.

The scientific community believes it has determined how long such parent elements will last. For example, it claims that rubidium-87 will change into strontium-87 within 48.8 billion years. Uranium-238 will change into lead-

[266] Ross, *Creation and Time*, 94–95; Ross, *Navigating Genesis*, 161.

[267] Marcos Ross, *The Heavens and the Earth: Excursions in Earth and Space Science* (Dubuque, IA: Kendal Hunt, 2015), 171.

[268] Ham and Hodge, *A Flood of Evidence*, 64.

206 in about 4.6 billion years, and potassium-40 will change into argon-40 in 1.3 billion years.[269] Marcos Ross opined, "the rate values [of converting the parent element to the daughter element] are rather cumbersome to use, so scientists often convert these into a different value, call a half-life."[270] One could say that the half-life of uranium-235 to lead is about 4.6 billion years, and the half-life of carbon-14 to nitrogen-14 is about 5,700 years.

Six-day creationists revealed that radioactive dating methods are not reliable. For example, a Hawaiian Island in 1800–1801 produced Hualalai basalt from the lava flow. Andrew Snelling noted that when secular scientists tested for the potassium and argon ratio in the 1960s and reported in secular scientific literature, the basalt age was between 1.32 and 1.76 million years old.[271] The radiometric dating method was false by a factor of 10,000. Similar results happened when the rock formed from Mount St. Helens in 1986 returned with a date of up to 2.8 million years old.[272] The six-day creationists have cast doubt upon the reliability of the radiometric dating method. In other words, the scientific methodology of advocates of PDAC that they claim affirms an old earth cannot detect what it claims.

In favor of young-earth creationism, six-day creationists claim they have scientific evidence that affirms a young earth. First, they found carbon-14 in diamonds that are claimed to be 1–3 billion years old.[273] Yet the half-life of carbon-14 is about 5,700 years. The following comments are from Jim Mason, an experimental nuclear physicist, regarding carbon-14:

> Machines have made it possible to detect much smaller amounts of [carbon-14], but eventually there is so little [carbon-14] left in the

---

[269] Ross, et al., *The Heavens and the Earth*, 171.

[270] Ross, et al., *The Heavens and the Earth*, 170.

[271] Andrew Snelling, "Excess Argo: The 'Archilles' Heel' of Potassium-Argon and Argo-Argon Dating of Volcanic Rocks," *Impact*, 1999, https://www.icr.org/article/excess-argon-archilles-heel-potassium-argon-dating.

[272] Ham and Hodge, *A Flood of Evidence*, 68.

[273] Don DeYoung, *Thousands . . . Not Billions: Challenging the Icon of Evolution, Questioning the Age of the Earth* (Green Forest, AR: Master Books, 2005), 56–57.

sample that it becomes undetectable even by these sophisticated machines. It would take about 15.6 half-lives of [carbon-14] (about 90,000 years) for the [carbon-14] in a sample to decay to the point where these modern machines could no longer detect [carbon-14].[274]

Mason disclosed that carbon-14 should not be present in diamonds if diamonds are billions of years old. Only diamonds that are thousands of years old would have detectable carbon-14. For advocates of PDAC, the presence of carbon-14 is a problem but not for six-day creationists. The carbon-14 dating indicates that the diamonds are thousands of years old and consistent with Genesis' timeline.

Six-day creationists have also noted that Mary Schweitzer, a paleontologist who is not a young-earth creationist, discovered soft tissue in dinosaur bones.[275] Although the dinosaur was supposedly sixty-eight million years old, she found red blood cells and heme. Andrew Snelling remarked, "is it remotely plausible that blood vessels, cells, and protein fragments can exist largely intact over 68 million years?"[276] Andrew Snelling added, "Heme is a part of hemoglobin, the protein that carries oxygen in the blood and gives red blood cells their color."[277] The discovery of red blood cells is contrary to the prevailing scientific community. The red blood cells should not exist if the dinosaur bone is millions of years old.

Mark Armitage and Jim Solliday reported on other dinosaur remains that revealed soft tissue, this time from a duckbill dinosaur cartilage, which contained vascular veins, venule valves, and nerve fibers.[278] What made this

---

[274] Jim Mason, "Radiometic Dating," in *Evolution's Achilles' Heels,* ed. Robert Carter (Powder Springs, GA: Creation Book, 2014), 207.

[275] Helen Fields, "Dinosaur Shocker," *Smithsonian Magazine,* 2006, https://www.smithsonianmag.com/ science-nature/dinosaur-shocker-115306469/.

[276] Andrew Snelling et al., "10 Best Evidences from Science that Confirm a Young Earth," in *Best Evidences* (Hebron, KY: Answers in Genesis, 2013), 28–31.

[277] Snelling et al., "10 Best Evidences from Science," 28–31.

[278] Mark Armitage and Jim Solliday, "UV Autofluorescence Microscopy of Dinosaur Bone Reveals Encapsulation of Blood Clots within Vessel Canals," *Microscopy*

discovery of soft tissue engaging was that the microscopy showed extensive clotting in most vessel canals. Armitage and Solliday contributed the clotting to asphyxia while drowning.[279] Asphyxia aligns with a Genesis account that all air-breathing, land-dwelling animals not on the ark died of drowning (asphyxia).

Armitage and Solliday added that since 2009, "19 museum bones specimens, including 15 dinosaur bones, selected from the paleontology collection at the University of Alberta, showed that all specimens yielded soft tissues."[280] For PDAC, finding soft tissue in dinosaur bones is a problem. The continuous finding of soft tissue casts doubt upon an old earth. The preservation of soft tissue in bones does not seem possible.

For SDC, it is not a problem to find soft tissue in dinosaur bones. Six-day creationists insist that since God sent a global flood about four thousand years ago, preserved soft tissue in dinosaur bones is entirely possible and seems to be the normal discovery of the paleontology community. Consistent with the biblical timeline of the genealogies of Genesis 5 and Genesis 11, preserved dinosaur tissue aligns with the Scripture. Six-day creationists continue to assert the congruity of observational science and the Scripture.

Summary of Radiometric Dating

Advocates of both PDAC and SDC look at the same evidence. For PDAC, the radiometric dating points to a universe and earth that are billions of years. They interpret the data according to their view and conclude the earth is quite ancient. For SDC, the radiometric dating produced a different interpretation that was more aligned with a universe and earth of thousands

---

Today, 28, no. 5 (2020): 30–38,
https://www.cambridge.org/core/journals/microscopy-today/article/uv-autofluorescence-microscopy-of-dinosaur-bone-reveals-encapsulation-of-blood-clots-within-vessel-canals/8762E671960898DAC303973A5A2A93F6#.

[279] Armitage and Solliday, "UV Autofluorescence," 30.

[280] Armitage and Solliday, "UV Autofluorescence," 30.

of years. Six-day creationists have raised questions about radiometric dating reliability.

Something is scientifically inaccurate with the supposed reliable dating method when submitted geological rock with a known date of formation (e.g., Hualalai basalt formed during 1800–1801 and rock from Mount St. Helen formed in 1986) returns with a geological date of millions of years old. Six-day creationists have presented counter-evidences, such as carbon-14 in diamonds and soft tissue regularly discovered in dinosaur bones, as evidence that the earth is thousands, not billions of years old.

SUMMARY OF THE ANALYSIS AND EVALUATION OF HUGH ROSS'S DAY-AGE CREATIONISM

The division between PDAC and SDC regarding the age of the world is vast. Advocates of PDAC claim the world began billions of years ago, but the advocates SDC assert the world began thousands of years ago. The debate is about more than just interpreting scientific evidence. The most important part of the debate is about the presuppositions of each group and their biblical hermeneutics. The PDAC view affirms the equality of general and special revelation in theory. Simultaneously, in practice, the PDAC view elevates its understanding of general revelation above special revelation, which means that the prevailing scientific discovery of advocates of PDAC will influence the interpretation of many evangelical theologians regarding the creation event and Noah's flood.

Advocates of PDAC conclude that the Scriptures are consistent with the widespread view that the universe is billions of years old. In other words, the creation event did not happen over six twenty-four-hour periods of time rather over billions of years. Their scientists and theologians read Genesis as a historical narrative and interpret it according to scientific consensus.

The SDC view affirms the supremacy of special revelation over general revelation, which means their scientists and theologians give primacy to the

Scriptures as authoritative when it comes to the universe's origins. In their estimation, interpretations of scientific discoveries fit within their understanding of the creation event. Six-day creationists read Genesis, primarily, as a historical narrative. For them, since God was present when the universe began and cannot lie, His explanation of its origins is final. Six-day creationists affirm God created the universe over six twenty-four-hour days. The disparity between the presuppositions of the advocates of PDAC and SDC provides the primary reason for their vastly different age of the universe.

PDAC affirms the universe is billions of years old based upon two streams of evidence. First, the starlight from the most distant stars is billions of light-years away. Second, radiometric measurement dates rocks at the lower levels of the sedimentary layers at billions of years. SDC has provided two potential solutions to the distant starlight. First, Humphreys proposed two gravitational time dilations that could have occurred. One gravitational time dilation occurred at creation, and the other gravitational time dilation occurred during Noah's flood. Lisle proposed that the one-way rate of the speed of light could be instantaneous and allowed light on day four to arrive virtually instantaneously to earth on the same day.

Six-day creationists disclosed problems with the reliability of radiometric dating. They presented observational evidence of carbon-14 in diamonds and soft tissue in numerous dinosaur bones. In their opinion, both pieces of evidence point to an earth that is thousands rather that billions of years old.

The theological and scientific interpretations of PDAC and SDC are different. Each group determines the earth's age based upon their reading of the Scriptures and what role scientific observations play. Both claim that the Bible is the inspired word of God. However, only one interpretation of the Scriptures is accurate, and only one understanding of the creation event is rooted in the Scriptures.

Chapter 4

The Contribution of Biblical Theology to the Topic of the Earth's Age

Ross's progressive day-age creationism view (PDAC) and the six-day creationism view (SDC) interprets Genesis's creation event differently. The PDAC view asserts that the earth is billions of years old, while the SDC view claim the earth is thousands of years old.[281] Several scriptural passages address the topic of creation. An intertextual analysis of these selected passages will help clarify which of the two views best reflects the Scriptures' depiction of creation. The following sections interpret selected passages from the Old and New Testament that refer to the creation event.

SELECTED PASSAGES IN THE PENTATEUCH

Confirmation of Adam's Creation (Genesis 5:1–2)

The author of Genesis affirms that God created Adam, a man made in the image and likeness of God, on day six. Adam became the father of Seth, a man made in the image and likeness of Adam. This parallel echoes the creation account. Allen Ross states that Genesis 5 "begins with a reiteration of the creation of man in the likeness of God."[282] The divine-human authors

---

[281] Ken Ham, "Young-Earth Creationism," in *Four Views on Creation, Evolution, and Intelligent Design*, ed. J. B. Stump (Grand Rapids: Zondervan, 2017), 24; Hugh Ross, "Old-Earth (Progressive) Creationism," in *Four Views on Creation, Evolution, and Intelligent Design*, eds. J. B. Stump (Grand Rapids: Zondervan, 2017), 87.

[282] Allen P. Ross, "Genesis," vol. 1, *The Bible Knowledge Commentary: An Exposition of the Scriptures*, ed. J. F. Walvoord and R. B. Zuck (Wheaton, IL: Victor Books, 1985), 35.

of Genesis reemphasized Adam's historical reliability and referred to the creation account of Adam in Genesis 1:26–28. Additionally, Gordon Wenham argued that Genesis 5 began as a short legal document referencing the historical account of Adam's creation.[283] Kenneth Matthews furthermore claimed that the author of Genesis used a written source to compile the genealogies of Genesis 5.[284] The word תּוֹלְדוֹת, translated *account*, *book*, or *record*, suggested that perhaps a brief theological and biographical source existed that the author compiled into the final form of the book of Genesis.[285] Therefore, the Scriptures clearly reveal only a real person named Adam, who God created on day six, could produce physical offspring eventually traced to Noah.

Honor the Seventh Day (Exodus 20:8–11)

After Moses led the nation of Israel out of slavery from Egypt, Yahweh told Moses to climb to the top of Mount Sinai. There He would give Moses's instructions for the nation to obey. Ed Hindson and Gary Yates remarked, "The Mosaic covenant explains how the nation [of Israel] was to be organized under God's kingship." [286] While on the mountain, Yahweh gave Moses the Ten Commandments.

In Exodus 20:8, Yahweh commanded Israel to keep the Sabbath day. The Sabbath day was the seventh day of the week, based upon the six-day creation week. Daniel Akin observed that God based the Sabbath day commandment firmly on the creation week's order.[287]Yahweh commanded Israel to set apart the Sabbath day from the other six days. The reason Yahweh set apart the

---

[283] Gordon J. Wenham, *Genesis 1–15*, *Word Biblical Commentary*, vol. 1 (Dallas: Word, Incorporated, 1987), 125.

[284] K. A. Mathews, *Genesis 1–11:26*, *The New American Commentary*, vol. 1A (Nashville: Broadman & Holman, 1996), 306.

[285] Matthews, *Genesis 1–11:26*, 306.

[286] Ed Hindson and Gary Yates, eds., *The Essence of the Old Testament: A Survey* (Nashville: B&H, 2012), 78–79.

[287] W. A. Criswell, et al., eds., *Believer's Study Bible*, electronic ed. (Nashville: Thomas Nelson, 1991), Exodus 20:11.

seventh day was that He created the world in six days and rested on the seventh day.

God modeled the six-day workweek and then rested, establishing an example for the nation of Israel to follow. Ken Ham commented, "The commandment makes no sense if the days are not literal in verse 11 as they are in verses 8–10."[288] Yahweh provided clarity at the end of verse 11 explaining that within a timeframe of six days, He created everything that exists. On day seven, He rested so Israel also must rest on the seventh day.

Douglas Stuart also noted that by working six days and resting one day, God modeled in His creation event a pattern for humankind:

> There could hardly be a stronger model for keeping the Sabbath than that of God himself. And there could hardly be a more impressive precedent within history (and soon enough, within the Scripture itself) than the creation account of Gen 1, which purposely describes creation in terms of six days of labor and one day of rest. God's model in this matter obviates all objections from anyone that he or she "doesn't need to take a day off" since God could hardly wear himself out.[289]

Stuart asserted that the Sabbath day makes no sense apart from six literal days of creation. For this reason, Israel was to work for six days and then rest for one day.

Moses, the author of Exodus, reinforced this understanding of the length of time of the creation week. Just as each day in the creation week was twenty-four hours, the basis for the Sabbath day commandment rested upon a literal understanding of the word יוֹם. Ham added, "to add millions of years into Genesis 1 . . . they [millions of years] can't be inserted into each of the days

---

[288] Ham, *Four Views*, 21–22.
[289] Douglas K. Stuart, *Exodus, The New American Commentary*, vol. 2 (Nashville: Broadman & Holman, 2006), 459–60.

or between the days or before the days, for the verse says 'in six days the LORD made the heavens and the earth, the sea, and *all that is in them* (emphasis Ham).'"[290] Moses, the author of Exodus, would not allow for an insertion of billions of years into the passage because his reference point was the seven-day week, grounded in the creation account. The God-ordained work week only makes sense with the pattern of a six-day creation event.[291]

Yahweh created the world in six days. Moses wanted Israel to know that those six days were the basis for the Sabbath day rest. The interpretation that best describes the length of each day of creation and the foundational pattern for the Sabbath day is a literal day of twenty-four hours.

Summary of Passages in the Pentateuch

In Genesis 5:1–2, the author confirmed the historicity of Adam. God created Adam on the sixth day and made him in His likeness. The author referred to Genesis 1:26–28, grounding the textual argument in the creation event. In the Scriptures, Moses clearly affirmed that Adam was a real person whom God created on the sixth day. The aforementioned passages underscore the historical reliability of a literal six-day creation.

In Exodus 20:8–11, Exodus's author revealed that God developed from His creation week the Sabbath day concept. In six days, He created the earth and then rested on the seventh day. Likewise, the nation of Israel would work for six days and then rest on the seventh day. The six days could only be twenty-four days, or Israel would not have understood the command to rest on the seventh day. Moses reinforced the six days of creation as literal days by means of the Sabbath day command.

---

[290] Ham, *Four Views*, 22.

[291] Robert V. McCabe, "A Critique of the Framework Interpretation of the Creation Week," in *Coming to Grips with Genesis: Biblical Authority and the Age of the Earth*, eds., Terry Mortenson and Thane H. Ury (Green Forest, AR: Master Books, 2008), 225–28.

SELECTED PASSAGES IN THE WRITINGS

Echoes of Adam's Origin (Job 33:4)

In the book of Job, Job suffered physically under God's care because He had allowed Satan to attack Job. Job believed he was innocent and challenged God to judge him honestly, even suggesting appropriate punishments for various sins.[292] Elihu reprimanded Job by giving various speeches. In one speech (Job 33:4), Elihu reminded Job that his existence is due to the Holy Spirit's work and created him and breathed into him.

Echoing back to Adam's original creation (Gen. 2:17), where the Holy Spirit involved Himself and gave life to Adam, all humanity existed because God created humankind.[293] Job had already affirmed the belief that God created him (Job 31:15). Elihu reminded Job that, just like at the beginning of creation, when God directly created Adam, God controlled all other events. So too, at the beginning of Job's life, God controlled all events. Job should trust God, the Creator of humanity because God knows his sixth-day creation better than anyone else.

Only God Can Describe the Creation Event (Job 38:4–6)

In Job 38:4–6, Job encountered Yahweh in a whirlwind. The context was Job clinging to his belief that his suffering did not match with his sins. He was seeking answers from Yahweh.

Yahweh asked him a series of questions that Job could not answer. One of those questions asked by Yahweh to Job was "where were you when I laid the foundation of the earth?" Bruce Waltke argued, "The creation of the 'earth' described in Job 38:4–7 can be harmonized best with the creation of

---

[292] Hindson and Yates, *The Essence of the Old Testament*, 244.
[293] Roy B. Zuck, "Job," *The Bible Knowledge Commentary: An Exposition of the Scriptures*, vol. 1, eds. J. F. Walvoord and R. B. Zuck (Wheaton, IL: Victor Books, 1985), 757.

the dry land called 'Earth' on the third day as described in Genesis 1:9–10."[294] Yahweh brought Job back to the creation event.

Robert Alden added, "Not even Adam and Eve, the first couple, were present at this inaugural event [the laying of the foundation of the earth] that marked our planet's birth. Job could not answer because he was not there and could not know."[295] Job could not answer the various questions God proposed because Job was not present at the creation of the world.

Only God could describe what happened and how it happened. God reminded Job that his knowledge was limited, so he ought to trust Yahweh because He is the Creator. His ways are complicated but dependable.

The Creator of Heaven and Earth (Psalm 24:1–2 and 33:6–7)

The Psalmist reminded his readers that God founded the earth upon the seas as a reference to the third day of creation. God created the heavens and all that dwells in them by speaking them into existence. Ironside commented on the relationship between the word of the Lord, the Bible, and the scientific observations of men:

> These are two witnesses to God. Creation and the Bible are both from the same source. Some people talk about the disagreement between the Bible and science. There is no disagreement between the Bible and science, that is, real science. Science consists of an orderly arrangement of proven facts explaining the universe, but where you simply get a lot of hypotheses that have never been proven that is not real science. Some of these may often be in conflict with the Bible, but never true science; because true science is simply the explanation of the physical universe, and the God who inspired the

[294] Bruce Waltke, "The Creation Account in Genesis 1:1–3, Part II: The Restitution Theory," *Bibliotheca Sacra* 132, no. 526 (1975): 144.

[295] Robert L. Alden, *Job, The New American Commentary*, vol. 11 (Nashville: Broadman & Holman Publishers, 1993), 370.

Bible made the universe. In this section of the Psalm you find the work and the Word of God testifying to His perfection.[296]

Ironside captured the disagreement between advocates of PDAC and SDC. SDC looks to align science with the Scriptures, but PDAC looks to align the Scriptures with consensus science. The Psalmist finds no conflict between the word of the Lord and His creation. Both point to the same author: Yahweh. Psalm 24 and 33 affirm that God created, and the way He created aligns with the Genesis 1 account.

Who is Wisdom at Creation? (Proverbs 8:22–31)

The author of Proverbs asserted in these verses that before God created anything, only He existed. The author presented literary connections between Proverbs and Genesis. For example, concepts like the beginning and earliest times (8:23) connect to the first day of creation. The description of mountains and hills (8:25) and an earth with no vegetation (8:26) harken back to the third day of creation. The literary picture of the first dust (8:26) echoes God creating Adam on the sixth day. The literary connection to Genesis in Proverbs 8:22–31 is self-evident.

Matthew Mcaffee provided nine "lexical links between these two texts [that are quite] . . . remarkable when taken as a whole."[297] Through the personification of God's attribute of wisdom, the author of Proverbs reinforced the historicity of the creation narrative in Genesis. God created the world, and the literary allusions to the days of creation in Proverbs reinforced the event's historicity—God created the world in six literal days.

---

[296] H. A. Ironside, *Studies on Book One of the Psalms* (Neptune, NJ: Loizeaux Brothers, 1952), 197.

[297] Matthew Mcaffee, "Creation and the Role of Wisdom in Proverbs 8: What Can We Learn?" *Southeastern Theological Review* 10, no. 2 (2019): 55.

Summary of Passages in the Writings

The writings of the Old Testament emphasize that God created the earth in six literal days. In Job 33:4, the author echoed Adam's creation, affirming that the Holy Spirit gave life to Adam. In Job 38:4–6, God reminded Job that he was not present when God formed the earth on the third day. In Psalm 24:1–2, a reference to the third day of creation, the author affirmed that God formed the seas. Psalm 33:6–7 aligns with the creation account of Genesis 1. Proverbs revisited the creation event by personifying God's attribute of wisdom (8:22–31). The beginning of creation included the formation of the earth, dry land, plants, and humans. The Old Testament writings do not hint that billions of years had elapsed during the creation week. On the contrary, the Old Testament authors affirm the plain meaning of each day of creation. They support that God created the entire universe in six twenty-four-hour days.

SELECTED PASSAGES IN THE PROPHETS

Yahweh Created the Heavens and the Earth (Isaiah 45:12, 18)

Isaiah affirmed two times within six verses that God created the earth as well as the heavens and the earth (45:12, 18). Gap theorists, such as C. I. Scofield, asserted that Isaiah 45:18 gives evidence of a gap between Genesis 1:1 and 1:2.[298] He argued that Isaiah declared God did not create the earth as תֹּהוּ (uninhabitable). However, in Genesis 1:2, the earth was תֹּהוּ. Scofield concluded that God created the earth perfect but later judged it between Genesis 1:1 and 1:2 in connection to Satan's rebellion against the Lord (Isa. 14:12, Ezek. 28:12). According to this view, an indefinite time (millions or billions of years) elapsed in Genesis 1:2. As a result, the Spirit of God hovered over the face of the waters to recreate the תֹּהוּ earth.

---

[298] C. I. Scofield, *Scofield Study Bible* (New York: Oxford University, 1967), 973.

Weston Fields presented three arguments against the gap theory reading of Isaiah 45:18. First, one cannot assume that תֹהוּ has the same meaning in Genesis as it does in Isaiah when it has a range of meanings (e.g., formlessness, confusion, unreality, and emptiness).[299] Second, Isaiah 45:18 is better translated, "He did not create it *to be* a waste, for inhabiting, he formed it." Third, the traditional six-day creationist view is that the state of the earth in Genesis 1:2 was *merely a step* in creating the earth as the habitation of humanity.[300] Thus, according to Fields, "when God was done with his creation work in relationship to the earth, it was not תֹהוּ. It was תֹהוּ in the purest sense only for a short period of time on the first day of creation."[301] Edward Young concurred with Fields:

> [T]hat the prophet's language has nothing to do with Genesis 1:2, for if it did the prophet would have been able to see that in the beginning the earth was a *tohu* (desolation). This, however, is a misunderstanding of the language. Isaiah does not deny that the earth was once a *tohu*; his point is that the Lord did not create the earth to be a *tohu*, for an earth of *tohu* is one that cannot be inhabited, and has not fulfilled the purpose for which it was created. The purpose rather was that the earth might be inhabited.[302]

Young reiterated God's original intention of the final product of earth—a habitation for humans. Young argued that Isaiah did not endorse the gap theory as espoused by Scofield. The condition of the earth at creation was תֹהוּ, but God did not intend for the earth to stay in that condition. Instead,

---

[299] Francis Brown, Samuel Rolles Driver, and Charles Augustus Briggs, *Enhanced Brown-Driver-Briggs Hebrew and English Lexicon* (Oxford: Clarendon, 1977), 1062.

[300] Weston Fields, *Unformed and Unfilled: A Critique of the Gap Theory* (Greenfield, AR: Master Books, 2005), 123.

[301] Fields, *Unformed and Unfilled*, 123.

[302] Edward Young, *The Book of Isaiah, Chapters 40–66*, vol. 3 (Grand Rapids: Eerdmans, 1972), 211.

God created it to be habitable for plants, animals, and humans. At the conclusion of the sixth literal day of creation, the earth was fully formed and fully filled.

Yahweh, Creator of Heavens, Earth, and Humanity (Zechariah 12:1)

Before Zechariah proclaimed the events that would lead to the second coming, he assured his readers that this prophetic word came from their Creator.[303] Carroll Stuhlmueller noted that Zechariah 12:1 is "a fragment of a hymn, honoring Yahweh the Creator."[304] The reliability of the prophetic word is based upon the reliability of the Creator God. The postexilic prophet affirmed the historicity of the creation event.

First, the textual evidence is Zechariah's statement that his prophecy came from the God who "lays the foundation of the earth." The phrase "lays the foundation of the earth" refers back to day three of creation, where God said, "let the dry land appear" (Gen. 1:9). Second, Zechariah states his prophecy is from the God who "forms the spirit of man within him," referring to day six of creation, where God "breathed into his [Adam's] nostrils the breath of life, and man became a living being" (Gen. 2:7).

According to Zechariah, the day of the Lord is inevitable because the word came from their Creator.[305] This is the same God who, in six literal days, created the world. Zechariah affirms the reliability of God's word by referring to the Genesis creation account.

Roy Gingrich added that Yahweh is "the faithful, covenant-keeping God who is *certain* to fulfill His promises ... the Creator-Sustainer God."[306] Within one verse (12:1), Zechariah highlighted God's creation of day three and day

---

[303] Warren W. Wiersbe, *With the Word Bible Commentary* (Nashville: Thomas Nelson, 1991), The Great Tribulation.

[304] Carroll Stuhlmueller, *Rebuilding with Hope: A Commentary on the Books of Haggai and Zechariah, International Theological Commentary* (Grand Rapids: Eerdmans, 1988), 144.

[305] E. Ray Clendenen, "The Minor Prophets," in *Holman Concise Bible Commentary*, ed. David S. Dockery (Nashville, TN: Broadman & Holman, 1998), 389.

[306] Roy E. Gingrich, *The Books of Haggai and Zechariah* (Memphis: Riverside, 1999), 52.

six. Zechariah supports the creation account that God created the world in six days and contradicts the PDAC interpretation.

Summary of Selected Passages in the Prophets

The echoes of God forming Adam from dust in Genesis 2:7 reverberated. The author in Isaiah 40:26 confirmed that God created the stars, an allusion to day four.

Isaiah's writings leave no room for Scofield's purported gap between Genesis 1:1 and 1:2. Instead, Isaiah argued that God did not create the earth as uninhabitable (תֹהוּ) but habitable for all life on earth. Zechariah persisted in affirming the heavens and the earth belong to Yahweh because He created them. Finally, as mentioned previously, Zechariah alluded to day three and day six creation events—the creation of the land and the formation of Adam.

SELECTED PASSAGES IN THE GOSPELS AND ACTS

In the New Testament, the authors of the Gospels, Acts, and Epistles made clear their affirmation of the creation event.[307] God inspired the New Testament authors. Therefore, God presents the historicity of the creation event in the Scriptures. He is, according to these authors, the creator of the six-day twenty-four-hour creation event.

Creation of Adam Affirmed (Matthew 19:4, Mark 10:6)

Both Matthew and Mark recorded Jesus's response when some Pharisees asked Him if the Law of Moses permitted divorce. Jesus responded by quoting Genesis 1:27, thus grounding His theology in Genesis. Craig Bloomberg remarked, "Matthew reproduces the LXX, rendering verbatim [Genesis 1:27]. The reason for the quote apparently is to stress that God

---

[307] Henry M. Morris, *The Biblical Basis for Modern Science* (Green Forest, AR: Master Books, 2002), 17.

created the two genders for each other.[308] This writer would add that another emphasis in Matthew's and Mark's writings was the authority and historical reliability of Genesis.

Jesus supported His understanding of marriage from the creation event. Genesis 1:27, according to Jesus, was part of the six-day creation account. In this passage, Jesus did not hint at long periods of time between each day. Jesus's statements, thus, affirm the beginning of the creation as depicted in Genesis 1:1–2:25. Additionally, Jesus's understanding of the beginning of all included Adam's creation (and eventually Eve). John Nolland argued, "ἀπ' ἀρχῆς ('from [the] beginning') [in Matthew 19:4] is likely to be intended to echo the opening phrase of Gen. 1:1 and to anchor the quotation from Gen. 1:27 in the first creation account of Gen. 1.[309] Donald A. Hagner declared, "'καὶ εἶπεν,' [in Matthew 19:4] 'and he said,' the subject of the verb is probably to be understood not as Jesus (as at the beginning of v. 4) but as the Creator, who is regarded as speaking through Moses."[310] Hagner, thus, argued that Jesus asserted His role as Creator in this passage. Jesus authoritatively knew the beginning of the world included Adam and Eve's creation because He created them on the sixth day.

Jesus affirmed Adam's creation account was historically accurate, and the event took place at the beginning of the world. The beginning for Jesus included all of Genesis 1–2. Thus, the beginning included six twenty-four-hour days of creation and one twenty-four-hour day of rest. Terry Mortenson summarized the theology of Matthew 19:4 and Mark 10:6: "[Jesus] demonstrated His young-earth viewpoint in Mark 10:6, Mark 13:19–20, and Luke 11:50–51. When analyzed carefully, 'from the beginning of

---

[308] Craig L. Bloomberg, "Marriage, Divorce, Remarriage, and Celibacy: An Exegesis of Matthew 19:3–12," *Trinity Journal* 11, no. 2 (1990): 166.

[309] John Nolland, *The Gospel of Matthew: A Commentary on the Greek Text*, The New International Greek Testament Commentary (Grand Rapids: Eerdmans, 2005), 771.

[310] Donald A. Hagner, *Matthew 14–28*, *Word Biblical Commentary*, vol. 33B, (Dallas: Word, 1995), 548.

creation' in Mark 10:6 refers to the beginning of the whole creation, not just the creation of the first marriage on day 6 of Genesis 1:27–30."[311]

The Suffering of Humanity Since the Beginning (Mark 13:19)

Jesus spoke to His disciples about the end of time, the great tribulation, and His return. In Mark 13:17–18, Jesus warned that when people see the abomination of desolation as predicated in Daniel 9:27 and 12:11, the tribulation would increase in intensity. John Grassmick encapsulated Jesus's description of the tribulation:

> Jesus looked beyond A.D. 70 to the final Great Tribulation prior to the Second Advent. This is supported by these facts: (a) Mark 13:19 echoes Daniel 12:1, an end-time prophecy; (b) the words "never to be equaled again" indicate that another crisis will never be like this one; (c) "those days" link the "near" future with the "far" future (d) the days will be terminated.[312]

The days of the great tribulation will be marked by events that had never happened since creation and would never happen again. When one reads the book of Genesis, the only event compared to the great tribulation was Noah's flood. Jesus inferred that even Noah's flood was not as disastrous as the future great tribulation.

Jesus argued that there would be a beginning and end of human suffering. It began at creation and will end when the great tribulation occurs. Ross argued that nonhuman suffering began at the origin of the world, but human suffering began at Adam's rebellion in Genesis 3.[313] Jesus denied the PDAC

---

[311] Terry Mortenson, "Jesus, Evangelical Scholars, and the Age of the Earth," *The Master's Seminary Journal* 18, no. 1 (2007): 69.

[312] John D. Grassmick, "Mark," vol. 2, *The Bible Knowledge Commentary: An Exposition of the Scriptures*, eds. J. F. Walvoord and R. B. Zuck (Wheaton, IL: Victor Books, 1985), 170.

[313] Ross, *Four Views*, 86.

assertion of billions of years of suffering. Jesus's understanding of the beginning of the origin of the world included the creation event and the fall. He did not allow for billions of years to elapse between creation and the fall.

Mortenson assumed for the sake of the argument that, if PDAC was correct, then Jesus should have said that since Adam, no such suffering has ever existed because, according to PDAC's proponents, suffering existed before the creation of Adam. Instead, Mortenson remarked, "[Jesus's] choice of words reflects His belief that man was there at the beginning and human suffering commenced essentially at the beginning of creation, not billions of years after the beginning."[314] For Jesus, the beginning of the world commenced with six days of creation, and according to Genesis, these days were six twenty-four-hour periods of time.

The Foundation of the Earth (Luke 11:50–51)

The key phrase of Luke 11:50–51 is ἀπὸ καταβολῆς κόσμου, "since the foundation of the world." What was Jesus's definition of the foundation of the earth? Jesus stated that wicked individuals had shed the blood of prophets since the foundation of the earth. In verse 51, Jesus asserted that the death of Abel was part of the foundation of the world. Thus, Jesus's definition of the foundation of the earth included the creation event as well as the death of Abel.

Cain killed his brother, Abel, at the beginning of human history. Mortenson asserted, "The parallelism is clear: 'blood' in both verses, the two temporal phrases beginning with ἀπό ('from' or 'since'), and repetition of 'charged against this generation.' The parallelism strongly suggests that Jesus knew that Abel lived very near the foundation of the world."[315] Howard Marshall added, "καταβολή is 'foundation, beginning,' always used in the NT with ἀπό or πρό to refer to the beginning of the world (except Heb.

---

[314] Mortenson, "Jesus, Evangelical Scholars," 75.
[315] Mortenson, "Jesus, Evangelical Scholars," 78.

11:11)."[316] Jesus understood the foundation of the world to include the death of Abel. If the PDAC view were correct, the foundation of the world could not have included Abel because billions of years would have elapsed before he died. Contextually, Jesus's words best fit with the SDC view.

Jesus Created All Things at Creation (John 1:1–5)

John asserted that Jesus, the second person of the Trinity, was with the Father from the beginning. The Father did not create the Son because He is eternal. In verse 3, John declared that Jesus made all things. Andreas Köstenberger stated about Jesus the Creator of all: "John's contention, however, is that everything-that is, the κόσμος (kosmos, world) of 1:10 came into being through 'him,' that is, *Jesus*, God-made-flesh (the word διά [dia, through] conveys secondary agency on the part of the Son here."[317] Everything that Köstenberger stated that Jesus created took place at the six-day creation event.

D. A. Carson explained that John showed Jesus created all things, and without Him, nothing could exist. Carson wrote, "[John] simply insists, both positively and negatively, that the Word was God's Agent in the creation of all that exists. Positively, *Through him all things were made*; negatively, *without him nothing was made that has been made* (emphasis Carson)."[318] As the Creator of the world, Jesus brought everything into existence during the six-day creation event. The apostle John emphasized the creation event in the opening of his Gospel account.

When the reader compares John 1:1–5 with Genesis 1:1–31, one will see that Moses asserted that God created everything, and John argued that Jesus created everything. The parallelism equates Jesus with God by emphasizing

---

[316] I. Howard Marshall, *The Gospel of Luke: A Commentary on the Greek Text*, *The New International Greek Testament Commentary* (Exeter: Paternoster, 1978), 505.

[317] Andreas J. Köstenberger, *John, Baker Exegetical Commentary on the New Testament* (Grand Rapids: Baker, 2004), 29.

[318] D. A. Carson, *The Gospel According to John, The Pillar New Testament Commentary* (Grand Rapids: InterVarsity, 1991), 118.

His deity and His creative acts. John structured the opening of his Gospel to reaffirm that Jesus was present with the Father, and Jesus created, along with the Father, the entire universe.

John commented on the creation event. Although he did not write a commentary on every aspect of the creation event, he did affirm that Jesus brough everything into being, just like God the Father brought everything into being. According to Moses, the creation event took six twenty-four-hour days. John's commentary does not assert a different time frame such as the PDAC view. To make such a claim the PDAC view would have to provide contextual evidence of John 1:1–5 contradicting a six-day creation event. John did not allow for the possibility of billions of years in the creation account. John's writings contradict the PDAC view.

The Human Race Begins with Adam at Creation (Acts 17:24–28)

In Paul's sermon on Mars Hill, he affirmed God created the world and that the human race originated from Adam. Simon Kistemaker explained, "Without mentioning his [Paul's] source, he teaches the creation account of Genesis and states that God is man's Creator (Gen. 2:7). Furthermore, out of one man, Adam, God made every nation on this earth."[319] The entire human race started with Adam on day six. All people groups originated from the first couple, Adam and Eve. The fact that the entire human race started with one couple means people from Asia, Africa, Europe, Australia, and South America are not people from separate races, but one race descended from Adam and Eve. Contrary to evolutionary dogma that asserted black people were inferior to white people, the Bible argues for the equality of all people.[320]

---

[319] Simon J. Kistemaker and William Hendriksen, *Exposition of the Acts of the Apostles, New Testament Commentary*, vol. 17 (Grand Rapids: Baker, 2001), 634.

[320] Jerry Bergman, "Evolution, Racism, and the Scientific Vindication of Genesis," in *Searching for Adam: Genesis & the Truth About Man's Origin*, ed. Terry Mortenson (Greenfield, AR: Master Books, 2016), 378–79.

Since all people are equal because they are derived from Adam on day six, ontologically, they are equivalent. Kenneth Gangel added, commenting on Acts 17:26, "God decides not only how long a nation stays on the map, but also how far it will reach before it is sent into decline by God. *He* determined how far the Roman Empire, or the British Empire, or Hitler's Third Reich would go before it came to an end. This is what Paul had in mind."[321] According to Paul, the existence of all of humanity is due to God creating Adam on day six.

Summary of Passages in the Gospels and Acts

The gospels and Acts do not deviate from the creation account of Genesis. Instead, they affirm the historicity of the creation event. The gospels support the creation of Adam and Eve as part of the beginning of everything. Their marriage began at the same time. The foundations of the world include the slaying of the prophet Abel. Thus, billions of years did not transpire between Genesis 1:1 and 4:8.

According to the New Testament, Jesus created everything that existed during the beginning. The beginning is the six days of creation. The author of Acts argued that all of the human race began in Adam on day six. Each author confirmed his belief as described in Genesis that the six-day creation event was literal.

SELECTED PASSAGES IN THE PAULINE EPISTLES

Death Through Adam's Sin (Romans 5:12)

Paul commented that Adam's sin brought death into the world. Unlike the PDAC view that claims death and bloodshed have always existed for plants and animals, Paul remarked that when Adam disobeyed Yahweh, sin

---

[321] Kenneth O. Gangel, *Acts, Holman New Testament Commentary*, vol. 5 (Nashville, TN: Broadman & Holman, 1998), 290.

entered the world, which resulted in death.[322] One could argue that Paul specifically addresses only the death of humans in Romans 5:12. However, Paul also addresses the death of the rest of creation in Romans 8:19–23.

Paul argues that at some point after the six days of creation, Adam disobeyed God. Contextually, that position seems to be a reasonable interpretation. However, a few chapters later (Rom. 8:19–23), Paul commented on the curse's effects upon plants and animals. The only place in the Scriptures that God cursed plants, animals, and humans was in Genesis 3:14–20. Thus, when Paul commented on the effects of death upon humans in Romans 5:12, he relied upon Genesis 3:14–20 as the primary source to explain the curse upon all of creation. Romans 5 and 8 both affirm that Adam's sin initiated the death of humans as well as the death of the entire creation. Death, disease, and bloodshed entered all of God's very good creation at the same time, during the period that Jesus called (in Luke 11:50–51) ἀπὸ καταβολῆς κόσμου "since the foundation of the world."

In Romans 5:12, Paul also affirmed the historicity of Adam. When God breathed life into him, his existence began in that moment and on the sixth day of creation. John Whitmer opines, "God's penalty for sin was both spiritual and physical death (cf. Rom. 6:23, 7:13), and Adam and Eve and their descendants experienced both. But physical death, being an outward, visible experience, is in view in 5:12–21."[323] The historical reality of Adam and Eve's sin transferred to all their descendants. The reason people die is because of the curse of sin that originated with Adam (and Eve). Paul affirmed not only the historicity of Adam but also that God created Adam on day six. Paul emphasized the fall and its effects upon humanity. For Paul, the death of humanity began shortly after Adam's creation.

---

[322] Hugh Ross, *Creation and Time* (Colorado Springs: NavPress, 1994), 61.

[323] John A. Witmer, "Romans," vol. 2, *The Bible Knowledge Commentary: An Exposition of the Scriptures*, eds. J. F. Walvoord and R. B. Zuck (Wheaton, IL: Victor Books, 1985), 458.

Creation Awaits the Reversal of the Genesis Curse (Romans 8:19–23)

As previously discussed in relation to Romans 5:12, Paul described the effects of the Genesis 3 curse on humankind. In Romans 8:19–23, Paul explained the effects of the Genesis 3 curse on the rest of creation. The earth waits for God to release His curse upon it. Death, disease, and bloodshed exist because of Adam's sin. Thomas R. Schreiner remarks, "the text [of Romans 8:20] should be interpreted in terms of Gen. 3:17–19, where the ground is cursed because of human sin, and thereby does not fulfill its created purpose."[324] God designed creation for a purpose, yet because of Adam's sin, creation waits to serve its purpose again once God releases it from its curse.

The release of the curse will occur when God creates the new heavens and new earth (Rev. 22:3). Kenneth Boa and William Kruidenier add, "When the curse is lifted, the creation will once again be an Edenic environment suitable for the image-bearers of God to inhabit and to reflect the Creator's glory. At present, the creation reflects the curse of sin; when sin is finally removed from the children of God, the creation will spring forth in glory."[325] In Romans 8:19–23, Paul reinforced the trustworthiness of Genesis by affirming the historical reliability of Adam, Adam's rebellion against his Creator, and God's curse upon creation.

Reaffirmation of the Historical Adam (1 Corinthians 15:21–22)

Similar to Romans 5:12, Paul in 1 Corinthians 15:21–22 affirmed the historicity of Adam which began on day six of the creation event.[326] Paul commented on Genesis 3:17–19 that Adam's sin imputed sin to all people, and the result of that sin was death for all individuals.

---

[324] Thomas R. Schreiner, *Romans, Baker Exegetical Commentary on the New Testament*, vol. 6 (Grand Rapids: Baker, 1998), 436.

[325] Kenneth Boa and William Kruidenier, *Romans, Holman New Testament Commentary*, vol. 6 (Nashville, TN: Broadman & Holman Publishers, 2000), 257.

[326] David K. Lowery, "1 Corinthians," vol. 2, *The Bible Knowledge Commentary: An Exposition of the Scriptures*, eds. J. F. Walvoord and R. B. Zuck (Wheaton, IL: Victor Books, 1985), 543.

Anthony Thiselton notes, "Adam is, for Paul, both an individual and a corporate entity: he was what his Hebrew name signifies—'mankind.' The whole of mankind is viewed as originally existing in Adam."[327] Once more, Paul affirmed the historicity of Adam and implied his belief that God created Adam on day six.

Summary of Passages in the Pauline Epistles

Paul confirmed that all humanity has known God exists. The evidence began with the creation of humanity, which occurred on the sixth day of creation (Rom. 1:20). According to Paul, Adam was a real person, and Adam sinned against his Creator.

Nothing in Paul's writing indicates he believed anything other than the fact that God created Adam on day six and that each day of creation was a literal twenty-four-hour period. In Romans 5:12 and 8:19–23, Paul reiterated the continued effects of the historical Genesis 3 curse that followed the six days of creation that God placed upon humanity and the rest of creation.

SELECTED PASSAGES IN THE GENERAL EPISTLES

God the Son the Creator of the World (Hebrews 1:2, 11:3)

The author of Hebrews stated that God the Son created the world. By the command of His voice, He created the universe out of nothing (*ex nihilo*). Simon Kistemaker notes, "The writer of Hebrews immediately clarifies the term *all things* by saying that God made the universe through his Son. The phrase obviously refers to the creation account in the first chapters of Genesis."[328] Donald Guthrie remarks, "Science could not reject the idea *that the world was created by the word of God*, for this view does not rest upon a

---

[327] Anthony C. Thiselton, *The First Epistle to the Corinthians: A Commentary on the Greek Text, The New International Greek Testament Commentary* (Grand Rapids: Eerdmans, 2000), 1225.

[328] Simon J. Kistemaker and William Hendriksen, *Exposition of Hebrews, New Testament Commentary*, vol. 15 (Grand Rapids: Baker, 2001), 28.

scientific evaluation of the 'seen' facts. The writer recognizes that acceptance of a special creative act of God is possible only by faith. . . . But the words *By faith we understand* show that knowledge is not independent of faith."[329] The faith believers possess is not a blind faith without reliable testimony. Rather, it is faith resting upon the testimony of God through His revelation.

On the contrary, believers have the testimony of God in the Scriptures. God created the world, and "[t]he author [in Hebrews 11:3] denies the eternity of matter, a common theory then and now, and places God before the visible universe as many modern scientists now gladly do."[330] The author of Hebrews reinforces the Genesis creation account, which includes six twenty-four-hour days.

Affirmation of the Creation of the Earth and Noah's Flood (2 Peter 3:5–6)

Combating the mockers who denied the judgment of God by means of the second coming of Christ, Peter reminded his readers that there have always been scoffers. Some scoffers endorsed a form of uniformitarianism regarding the history of the earth. Gangel comments, "[uniformitarianism] is the view that the cosmic processes of the present and the future can be understood solely on the basis of how the cosmos has operated in the past."[331] The mockers rejected miracles and claimed that things have always been the same. According to the scoffers, Jesus was not returning. Therefore, they did not fear His coming judgment. Gangel adds, "So [the mockers] referred to our fathers, that is, Old Testament patriarchs and to the beginning

---

[329] Donald Guthrie, *Hebrews: An Introduction and Commentary, Tyndale New Testament Commentaries*, vol. 15 (Downers Grove, IL: InterVarsity, 1983), 229.

[330] A. T. Robertson, *Word Pictures in the New Testament* (Nashville, TN: Broadman Press, 1933), Heb. 11:3.

[331] Kenneth O. Gangel, "2 Peter," vol. 2, *The Bible Knowledge Commentary: An Exposition of the Scriptures*, eds. J. F. Walvoord and R. B. Zuck (Wheaton, IL: Victor Books, 1985), 875.

of Creation. Since nothing has happened in all this time, mockers reasoned, why expect the Lord's return now?"[332]

Peter countered the mocker's allegation of uniformitarianism (God had not interfered with His creation since day six), asserted that God formed the earth out of the water (Gen. 1:2), and then used that same water to flood the entire earth in judgment. Peter Davids comments, "[Peter] moves on to the time of Noah and underlines the narrative point that these same waters that had been used by God to form the creation were used by God to destroy the creation."[333]

Davids's point was that the form of uniformitarianism that the mockers asserted was biblically incorrect. God had judged the earth through a world-wide deluge, Noah's flood, thus, disrupting the mocker's supposed view that nothing has changed since the earth's creation. Since God had judged the earth, the mockers could expect God to judge the earth again at the second coming.

Peter stated that the mockers appeared to have forgotten their biblical history. Davids comments, "the forgetting is not deliberate, as if the 'scoffers' were conscious that they were suppressing data, but it is a result of their desire that it be true that there is no evidence for a final judgment."[334] In other words, the scoffers wanted their viewpoint of uniformitarianism to be accurate. Thus, they had deliberately forgotten that God had already judged the earth with a global flood. Peter affirmed the historical reliability of creation (3:4).[335] Then he asserted the historicity of Noah's global flood (3:6). Noah's flood refuted the mockers' claim of uniformitarianism.

---

[332] Gangel, "2 Peter," 875.

[333] Peter H. Davids, *The Letters of 2 Peter and Jude*, The Pillar New Testament Commentary (Grand Rapids: Eerdmans, 2006), 270.

[334] Davids, *The Letters of 2 Peter and Jude*, 268.

[335] Michael Green, *2 Peter and Jude: An Introduction and Commentary*, Tyndale New Testament Commentaries, vol. 18 (Downers Grove, IL: InterVarsity, 1987), 152.

The mockers believed that since the creation event, God had not interfered with His creation in judgment. However, God had judged with water. Therefore, Christ will return to judge with fire (3:7).

Summary of Passages in the General Epistles

The General Epistles upheld that Jesus created the universe. Out of nothing, God created the earth. By God's word, the heavens and earth began. His creative power brought into existence what had not existed. The General Epistles also confirmed the historical reliability of Noah's flood. Peter refuted mockers who affirmed a form of uniformitarianism. The mockers believed that since day six, God had not interfered with His creation by means of judgment. However, God did enter His creation through the means of Noah's flood and disrupted the usual ebb and flow of life. The flood was a reminder to all that judgment would come again when Christ returns.

SELECTED PASSAGES IN THE APOCALYPSE OF JOHN

Creator of Heaven, the Earth, and the Seas (Revelation 10:6)

In the book of Revelation, John describes a scene during the seven-year tribulation period. He sees an angel give glory to God. The angel swears that God is the one who created the heaven, the earth, and the sea. G. K. Beale remarks, "The references to heaven, earth, and the sea, followed in each case by καὶ τὰ ἐν αὐτῇ ("and the things in it"), underscore the absolute sovereignty of God in creating *all* things."[336]

The textual evidence that supports John referring to the creation account is threefold. First, John's statement that God "who created heaven and the things in it," refers back to the opening statement of Genesis, where God says, "In the beginning, God created the heavens" (Gen. 1:1). John's account also refers to day four, where God filled the heavens with stars, sun, moon,

---

[336] G. K. Beale, *The Book of Revelation: A Commentary on the Greek Text*, The New International Greek Testament Commentary (Grand Rapids: Eerdmans, 1999), 538.

and planets (Gen. 1:14–19). Second, John's statement that God "created the earth and the things in it," refers back to the opening statement of Genesis, where God says, "In the beginning, God created . . . the earth" (Gen. 1:1). John's account also refers to days one through six, where God filled the earth with vegetation, animals, and humans (Gen. 1:9–13, 24–28). Third, John's statement that God "created the sea and the things in it," refers back to day five, where God says, "Let the waters teem with swarms of living creatures" (Gen. 1:20–23). Within one verse (Rev. 10:6), John highlights God's creation in general (Gen. 1:1), as well as His creative acts on days three, four, five, and six. According to John, as God was sovereign over the creation of the earth, so too will God reign sovereignly over the end.[337]

The Genesis Tree of Life and Curse (Revelation 22:2–3)

In the book of Revelation, John described the new heaven and earth. From Christ's throne originated a river, and on either side of the river was the tree of life. The tree of life reminds the reader of the garden of Eden. By means of this referencing, John connects the historicity of Genesis with the new creation.

John also observed the New Jerusalem and commented that the curse of Genesis 3 placed upon Adam, Eve, and nature no longer existed because God had removed it. Morris comments, "This [the removal of the curse] is the fulfillment of a prophecy (Zech. 14:11, 'there shall be no more curse'). Instead, *the throne of God and of the Lamb* are there."[338] Humanity can have full access to God because He has reversed the curse. Douglas Mangum adds, "In new Jerusalem, the curse is removed. Humanity once banished from God's presence at the invocation of the curse, but the curse's reversal renews humanity's access to God's presence."[339]

---

[337] Beale, *The Book of Revelation*, 538.

[338] Leon Morris, *Revelation: An Introduction and Commentary*, Tyndale New Testament Commentaries, vol. 20, (Downers Grove, IL: InterVarsity, 1987), 243.

[339] Douglas Mangum, ed., *Lexham Context Commentary: New Testament* (Bellingham, WA: Lexham Press, 2020), Eden Restored.

John takes the reader back to a literal tree of life in a literal garden of Eden that God created on day six for the literal humans, Adam and Eve. No more would humanity struggle with one another or nature. God would restore the new creation with the conditions of the garden of Eden. The sin nature as a result of Adam's rebellion in Genesis 3 that separated humanity from its Creator would no longer exist. Humanity is no longer separated from its Creator.

Summary of the Passages in the Apocalypse of John

The book of Revelation underscores God as Creator, highlighting His creative acts on days three through six. John also emphasizes at the end of his letter that God removes the curse of sin. In the new creation, God has placed the tree of life (a reminder of the creation event) in its midst. The Bible begins with the six days of creation and no sin, and the Bible ends with a new creation and no sin. Sin entered the world because of Adam's rebellion. Jesus provided a remedy for sin through His death and resurrection. The Creator in Genesis concluded His metanarrative for humanity in Revelation. The tree of life exists again, and the curse from Genesis 3 no longer exists.

SUMMARY OF THE SELECTED PASSAGES
IN THE OLD AND NEW TESTAMENT

The Old and New Testament writers affirmed a six-day creation, the historicity of Adam, the fall, and future redemption. The intertextual commentary of the Bible in selected passages from the Old and New Testaments reinforces the SDC view of the earth's age. In contrast, the PDAC view lacks a biblical and theological basis.

For example, the Pentateuch affirms the six days of creation and confirms each day was twenty-four hours. Also, the writings of Moses explain that God created the universe. The Pentateuch traces Noah's historical lineage back to Adam's lineage and asserts that Adam began the human race as described in Genesis 1 and 2.

The Writings did not hint that billions of years had elapsed during the creation week. On the contrary, the Writings support the plain meaning of each day of creation. The author of the book of Isaiah asserts that God created the universe and was the architect of Adam. Isaiah's writings do not support the gap theory. Instead, the prophet Isaiah argues that God did not create the earth as uninhabitable (תֹהוּ) but habitable for all life on earth.

The Gospels and Acts state that Jesus created the world. The Gospels further affirm that marriage began at the creation of Adam and Eve. Jesus, in the Gospels, asserts that Adam and Eve were historical individuals who God created during the creation event on the sixth day.

Through his Epistles, Paul affirms the historicity of the fall of Adam and that Adam's sin transferred to all his descendants. The result of the fall was death for humans, animals, and creation. The death of plants and animals did not begin billions of years before the death of Adam. On the contrary, death began when Adam disobeyed the Creator. Paul also argues that Jesus was the Creator and acknowledged the historical reliability of the creation event that Moses described in Genesis.

The General Epistles confirm the historicity of Noah's flood and deny a form of uniformitarianism. Even though Peter warns that mockers would doubt the judgment of the second coming, the proof that judgment was going to happen was that God had already judged the earth with a global deluge. The flood is a reminder that God can, at any time, interject Himself into His creation and disrupt what mockers consider the usual ebb and flow of life. The General Epistles also affirm that God created the world from nothing.

John, in the book of Revelation, also affirms that a day will come when God removes the Genesis 3 curse. At that time, God will create a new heaven, earth, and new Jerusalem. The removal of the curse marks the new creation as different from the old creation.

Six-day creationists find support for their view in both the Old and New Testaments. Collectively, the intertextual commentary of the other sixty-five books of the Bible points to a recent creation as depicted in Genesis. The PDAC is hard-pressed to make a case that the earth is billions of years old based upon the biblical authors' commentary. Instead, the commentary of the Old and New Testaments on the creation event affirms the earth is thousands of years old.

Chapter 5

The Doctrinal Significance
of Hugh Ross's Progressive
Day-Age Creationism

The Scriptures present a biblical theology that aligns closely with six-day creationism. By contrast, Ross presents a distorted theology that relies on a reinterpretation of the Scriptures. Ross's view, therefore, has doctrinal significance that alters the central teachings of the Christian faith. This chapter explores the theological implications of Ross's view in the areas of bibliology, theology proper, Christology, anthropology, soteriology, eschatology, and apologetics.

BIBLIOLOGY: AUTHORITY AND RELIABILITY
OF THE SCRIPTURES

God has revealed Himself in two ways: general and special revelation.[340] Through general revelation of nature, providence, and human conscience, God has revealed to all humanity that He exists. He has also revealed that He cares for His creation and that humanity knows a universal right and wrong exists.[341] General revelation is not sufficient to bring a person to the realization that Jesus is the Christ. Furthermore, general revelation does not

---

[340] Robert Lightner, *Handbook of Evangelical Theology* (Grand Rapids: Kregel, 1995), 11.

[341] James R. Estep Jr., Michael J. Anthony, and Gregg R. Allison, *A Theology for Christian Education* (Nashville: B&H, 2008), 75–76.

bring a person to the realization that he needs forgiveness of sins. People need a different kind of revelation that can make spiritually dead people come to life.[342]

Ross claimed that special revelation, the Bible, is superior to general revelation.[343] Ross, however, argues that general revelation "may be likened to a sixty-seventh book of the Bible . . . [and] God's truth [of special and general revelation] cannot be held as inferior or superior to another."[344] Moreover, in practice, Ross has devalued the role of special revelation and placed it below general revelation. For example, Ross insisted scientific observations and the interpretation of the Scriptures will lead to old-earth creationism, which is the thesis of PDAC.

Tim Chaffey and Jason Lisle note, "General revelation consists of information that has been available to all people throughout all time . . . [and] Ross incorrectly claims that modern scientific discoveries are a source of general revelation . . . [because] this information was not readily available to all people in history."[345] Ross's assertion implies that the church for roughly eighteen hundred years never understood Genesis 1–11 accurately. Ross claimed a general revelation from God that is universal for all to discern, but the church never understood. Implications from Ross's assertions is that God used Comte de Buffon, Jean Lamarck, and James Hutton, who sought to remove the need for Him, to reveal to His bride the correct understanding of Genesis 1–11. Ross's view stretches credulity to a breaking point.

---

[342] Michael Horton, *The Christian Faith* (Grand Rapids: Zondervan, 2011), 142. The writer affirms a mystery exists between the role of human responsibility and the work of the Holy Spirit in regeneration.

[343] Hugh Ross, *Creation and Time: A Biblical and Scientific Perspective on the Creation-Date Controversy* (Colorado Springs: NavPress, 1994), 57.

[344] Ross, *Creation and Time*, 56. This writer will explore in more detail in the section of apologetics the inconsistency of Ross's theology regarding his supposed claim of the superiority of special revelation over general revelation. He is not consistent, thus summarizing his view on special and general revelation requires more attention in the area of apologetics.

[345] Chaffey and Lisle, *Old Earth*, 76.

The grammatical interpretation of יוֹם within the context of the creation event can only have a meaning that is consistent with an approximate twenty-four-hour day. Ross imported a foreign idea of יוֹם into the contextual boundaries of the creation event because he was convinced general revelation has revealed that the earth is billions of years old. In practice, Ross has not placed special revelation over general revelation. He reversed the views even though in his writings he asserted that special revelation was superior to general. According to Ross, the reader needs the scientific community of Christians to interpret the Genesis account rightly. Yet, some of the scientific community of Christians cite the universe's age between fourteen and twenty billion years.[346] The cited age of the universe is quite a range and hardly precise.

From the days of the apostles until the eighteenth century, the Christian church has affirmed that the earth is thousands of years old.[347] Beginning in the eighteenth century, men such as Lamarck, Hutton, and Lyell asserted the earth was much older and challenged the prevailing view.[348] The majority of influential evangelical pastors and scholars concluded Lamarck, Hutton, and Lyell were correct and invented alternative theories of the creation event that could explain the supposed indisputable scientific evidence of an old earth within the plain-sense reading of Genesis 1:1–2:4.[349]

---

[346] Ross, *Creation and Time*, 101.

[347] Jeremy Sexton, "Evangelicalism's Search for Chronological Gaps in Genesis 5 and 11: A Historical, Hermeneutical, and Linguistic Critique," *Journal of the Evangelical Society* 61, no. 1 (2018): 5–25; Ernst Mayr, *What Evolution Is* (New York: Basic Books, 2001), 12; James Mook, "The Church Fathers on Genesis, the Flood, and the Age of the Earth," in *Coming to Grips with Genesis: Biblical Authority and the Age of the Earth*, ed. Terry Mortenson and Thane H. Ury (Green Forest, AR: Master Books, 2008), 23–52.

[348] Jean Lamarck, *Zoological Philosophy*, trans. Hugh Elliot (New York: Hafner 1964), xlii, 180, http://www.blc.arizona.edu/ courses/schaffer/449/Lamarck/Lamarck%20Zoological%20Philosophy.pdf; James Hutton, "Chapter 1" in the *Theory of the Earth*, vol. 1 (Royal Society of Edinburgh: Scotland, 1788), https://www.gutenberg.org/files/12861/12861-h/12861-h.htm; Charles Lyell, *Principles of Geology* (New York: D. Appleton & Co., 1830), 62–63.

[349] William Hanna, *Memoirs of Thomas Chalmers*, vol. 1 (Edmonston and Douglas: Scotland, 1867), 291; George Stanley Faber, *A Treatise on the Genius and Object of the*

Ross followed in the line of alternative theories of the creation event but now with more supposed scientific evidence than what Lamarck, Hutton, and Lyell possessed. He undermines the Scriptures reliability by directly asserting that the Scriptures do not claim God created the earth in six twenty-four-hour days.[350] Grammatically and historically, the Hebrew word יוֹם, surrounded by the contextual markers of עֶרֶב and בֹּקֶר, and qualified with a cardinal and ordinal number only, means a twenty-four-hour period.[351] Ross, however, attempted to redefine יוֹם to mean an indefinite, long period equaling billions of years.[352] Ross revealed his allegiance to scientific community interpretations over the church's grammatical and historical interpretations of Genesis.

According to Ross's view of the Scriptures, the author of Genesis 1 did not intend to communicate a literal six-day creation. According to Ross, the believer would be mistaken to assert that God created in six days. The believer must reject the grammar of Genesis 1, the intertextual commentary affirming a young-earth, and the historical interpretation supporting a young-earth view for the prevailing view of scientists who affirm an old earth. Based on Ross's hermeneutics, God's word cannot mean what it says unless first vetted by the scientific community. Apparently, then, Ross's final authority is not the Scriptures but his interpretation of scientific observations.

According to Ross's hermeneutic, one must wonder why he affirmed the resurrection of Jesus. This writer is not aware of any actual scientific observations by scientists proving that people can rise from the dead. This author does not doubt that Ross affirms the resurrection. Still, he does wonder how Ross can consistently affirm it when his interpretive grid of

---

Patriarchal, the Levitical, and Christian Dispensations, vol. 1 (Whitefish, MT: Kessinger, 2009), 116–18.

[350] Hugh Ross, *Genesis One: A Scientific Perspective*, 4th ed. (Covina, CA: Reasons to Believe, 2006), 25; Hugh Ross, *Creation and Time* (Colorado Springs: NavPress, 1994), 46.

[351] Chaffey and Lisle, *Old Earth*, 25–26.

[352] Ross, *Creation and Time*, 45, 95.

deferring to science as equal to the Scriptures have never verified that bodies could resurrect from the dead.[353]

A concern among six-day creationists is that a believer may begin to doubt the reliability of the remainder of the Bible if the believer cannot trust the opening chapters of Genesis.[354] Terry Mortenson captures the lack of humility that Ross and other PDAC proponents have displayed toward God's word: "True humility is demonstrated when we tremble at the Word of God. It is a great sadness to me that most inerrantist Bible scholars and theologians and other Christian leaders over the past 200 years have been in a very real sense trembling at the words of men (namely scientists), rather than trembling at the Word of God."[355] Mortenson affirms this author's concern that the PDAC view undermines the authority of the Scriptures. The God of Israel is the divine author of the Bible. He does not deceive (Num. 23:19). If what God authored in Genesis 1:1–2:4 is not what happened at the creation event and not according to the Bible's described timeline, believers cannot trust His words. God's character is the next theological doctrine at risk if Ross's view is correct.

Ross elevates his interpretations based upon scientific observation of the creation event above the clear, straightforward, and literal reading of Genesis. Ross's view undermines the authority and reliability of the Scriptures. If a believer follows Ross's view, his view could lead to doubting the other parts of the Scriptures. The final authority in all matters in faith, practice, history, and science should be the word of God.

---

[353] Ross, *Creation and Time*, 62.

[354] R. David Skinner, *Studies in Genesis 1–11: A Creation Commentary*, ed. Michael R. Spradlin (Collierville, TN: Innovo, 2018), xviii.

[355] Terry Mortenson, "Inerrancy and Biblical Authority: How and Why Old-Earth Inerrantists are Unintentionally Undermining Inerrancy," *Answers Research Journal* 13 (2012): 219.

THEOLOGY PROPER: THE SOVEREIGNTY
AND CHARACTER OF GOD

God's person and character, His attributes such as holiness and righteousness, are disclosed through the Scriptures.[356] God does what is right because what He does is consistent with His character, and His character is always righteous.[357] The believer and nonbeliever can rely upon the promise that God will not lie or deceive (Num. 23:19), thus, what God has authored through His word is also trustworthy. Since Ross asserts that the earth is billions of years old, however, in his mind the age of the earth that the church affirmed for over eighteen hundred years is not trustworthy.

Regarding God's sovereignty and providence, Ross affirmed what theologians call *middle knowledge*.[358] According to William Craig the middle knowledge theory affirms, "God possesses hypothetical knowledge of conditional future contingents, [such as] . . . counterfactuals."[359]

Counterfactuals are the outcome of the event that did not occur. For example, a car's driver speeds through a red light and hits another car entering the intersection. The counterfactual is the event's outcome in which the driver stops at the red light, and the accident does not occur. Instead, some other event happens.

Ross presented his reason for the sudden appearance of pain and suffering of nonhuman life from the beginning: "The current creation serves its purpose as the best possible realm in which God efficiently, rapidly, and

---

[356] Gordon R. Lewis and Bruce A. Demarest, *Integrative Theology*, vol. 2 (Grand Rapids: Zondervan, 1990), 177.

[357] Gerald Bray, *The Doctrine of God* (Downers Grove, IL: Intervarsity, 1993), 212–18.

[358] Ross, *Four Views*, 87. Based upon the author's understanding of the theology of middle knowledge, Ross affirmed either Dominican middle knowledge or a Molinian middle knowledge.

[359] William Lane Craig, "God Directs All Things on Behalf of a Molinist View of Providence," in *Four Views on Divine Providence,* eds. Stanley N. Gundry, and Dennis W. Jowers (Grand Rapids: Zondervan, 2011), 79–83.

permanently conquers evil and suffering while allowing free-will humans to participate in his redemptive process and plan."[360] Ross's affirmation of middle knowledge is his belief that God created the best possible world given all of the variables, including counterfactuals, that He knew would occur and then started the creation process. If Ross's view is correct, this author believes Ross has inadvertently presented God as less sovereign and more reactive to human choices. In other words, the beginning and the end are under God's control, but the abundance of human counterfactuals influenced how God would create His world. God made a world where the best chance for the gospel to flourish could occur, rather than God determining His own good pleasure by fashioning the world He wanted to create.

Based upon Ross's presupposition of deferring to the scientific community's consensus to interpret the Genesis creation account, instead of the grammatical-historical approach before the eighteenth century, readers could analyze all the Scriptures the same way. Readers could consult with the scientific community before interpreting the Scriptures instead of consulting with the theological community before interpreting philosophical science. Following the logical outgrowth of Ross's interpretation method, the straightforward Scriptures sections do not mean what the words communicate. Since the words do not mean what they communicate, the author behind the words is, at best, a poor communicator or, at worst, a deceiver. Either option casts doubt upon the character of God.

Humans are dependent upon their Creator.[361] They can only know their Creator if He reveals Himself in person, word, or action. Gordon Lewis and Bruce Demarest argue that the Scriptures affirm that "God made himself known through word and deed as a living word and active personal Spirit."[362]

---

[360] Ross, *Four Views*, 87.

[361] John S. Hammet, "Human Nature," in *A Theology for the Church*, ed., Daniel Akin (Nashville: B & H, 2007), 342.

[362] Lewis and Demarest, *Integrative Theology*, 184.

Ross's view implies that God veiled the supposed truth of the age of the earth in the early chapter of Genesis and waited for geologists-philosophers, who were hostile to Him, to exegete, not from the Scriptures, but from a cursed creation and declare earth billions of years old. Ross's view casts doubt upon God's character to communicate and casts doubt upon the Scriptures, which God said originates from Him.

According to Ross, the commentary from other Old Testament and New Testament books inaccurately affirmed that each day of creation was twenty-four hours with Adam and Eve's formation happening on the sixth day.[363] If readers follow Ross's assertions, God did not correctly communicate what He intended through His prophets', kings', and apostles' writings; He also failed to communicate through His Son.

Jesus claimed to create the earth, yet, inferring from Ross, Jesus failed to express the creation event that supposedly lasted billions of years. Instead, Jesus's words infer a recent creation of thousands of years.[364] The implications of Ross's arguments could contend that Jesus incorrectly taught that God created Adam and Eve at the beginning of creation (Matt. 19:4, Mark 10:6). Thus, the days of creation were not literal days even though the context leads to that conclusion.[365] If Ross's interpretation is correct, Jesus's teaching on the creation event included an error that reflects poorly upon God's character.

Finally, a logical conclusion from Ross's PDAC view is that men such as Hutton, Lyell, and Darwin, who sought to overturn the need for a Creator,

---

[363] Douglas K. Stuart, *Exodus, The New American Commentary*, vol. 2 (Nashville: Broadman & Holman, 2006), 459–60; John Nolland, *The Gospel of Matthew: A Commentary on the Greek Text, The New International Greek Testament Commentary* (Grand Rapids: Eerdmans, 2005), 771.

[364] This author is not aware of any passage in the Bible where Jesus's words could be interpreted to mean He affirmed an earth of billions of years old.

[365] Terry Mortenson, "Jesus, Evangelical Scholars, and the Age of the Earth," *The Master's Seminary Journal* 18, no. 1 (2007): 69.

exegeted the correct interpretation of Genesis 1–2:4.[366] Yet, these men did not exegete the inerrant word of God to discover the correct interpretation. To the contrary, they extrapolated the supposed correct meaning of the earth's age based upon the theory of uniformitarianism. Ross asserted his interpretations warranted him affirming his biblical and theological justification for an old earth.[367] The plain reading of the Scripture refutes Ross's theory.

Of course, Ross's intended interpretation may not be correct because future scientists could claim the earth is trillions of years old. In that case, Ross's current view would be erroneous right now. If the reader depends upon future scientific observations to validate the meaning of the Bible, the reader will ultimately never have confidence in the triune God through reading the Bible because scientists could overturn their old theories that currently support Ross's view.

If followed consistently, Ross's view leads to doubting God's ability to communicate clearly through His word and ultimately question His character. The readers of the Scripture will never know what God intended to communicate in Genesis because the reader will rely upon geologists-philosophers' prevailing view to clarify the earth's age. If the readers cannot trust the first few chapters of Genesis, they might doubt other passages of the Bible, which ultimately leads to questioning the trustworthiness of God's character. Ross's view carelessly creates an unnecessary doubt about the opening chapters of Genesis.

In contrast, the SDC view asserts a straightforward reading of Genesis 1:1–2:4.[368] The straightforward reading is consistent with the commentary

---

[366] Charles Darwin, *The Origin of Species* (New York: Signet Classics, New American Library, 1859, 2003), 4, 7, 14.

[367] Ross, *Creation and Time*, 53–72. Chapter six of his book is entitled "Theological Basis for Long Creation Days."

[368] Richard Davidson, "The Genesis Account of Origins," in *The Genesis Creation Account: And Its Reverberations in the Old Testament*, ed., G. A. Klingbeil (Berrien Springs, MI: Andrews University Press, 2015), 78; Chaffey and Lisle, *Old Earth Creationism*, 52; Ken Ham, *Six Days: The Age of the Earth and the Decline of the Church* (Green Forest, AR:

from selected Old and New Testament authors, especially Jesus. Until the eighteenth century, church history confirmed an almost universal belief that God created the world in six literal days. Only when men sought to overturn the creation narrative in the eighteenth century did believers doubt the creation account. The God of the Scriptures is reliable and trustworthy. He does not lie, deceive, or communicate in such a way that is inconsistent with His character. The creation account is true because God is true in all He does.

CHRISTOLOGY: THE TESTIMONY OF JESUS

Jesus affirmed the historicity of Adam and Eve, the fall, and the murder of Abel. All of those events happened at the beginning of the world. Jesus is the Creator, and He brought the universe into existence from nothing in six literal days. According to SDC, the world began between 6,000 and 10,000 years ago.[369]

According to Ross, Jesus created Adam between 6,000 to 135,000 years ago.[370] Ross has changed his creation date of Adam numerous times. However, he still holds that the earth is billions of years old. Thus, Ross concluded that Jesus promoted creating Adam and Eve on day six, however, Ross's understanding of day six is much further in the past than what the Scripture infers. According to Ross, God created them after billions of years of "mass extinction and speciation events occur[ring] with a semi-regular period of about thirty million years, timing that correlates with the up-and-down movement of the solar system relative to the galactic plane."[371]

Master Books, 2013), 65–77; Ken Ham, "Young-Earth Creationism," in *Four Views on Creation, Evolution, and Intelligent Design,* ed. J. B. Stump (Grand Rapids: Zondervan, 2017), 20; David Catchpoole, and Mark Harwood, "Ethics and Morality," in *Evolution's Achilles' Heels,* ed. Robert Carter (Powder Springs, GA: Creation Book, 2014), 235–60.

[369] Chaffey and Lisle, *Old Earth Creationism,* 23–24.

[370] Ross, *Creation and Time,* 140; Ross, *Genesis One,* 21; Hugh Ross, *Navigating Genesis: A Scientist's Journey through Genesis 1–11* (Covina, CA: RTB, 2014), 75.

[371] Ross, *Four Views,* 90.

Jesus stated that God created male and female at the beginning. This would seem to indicate the beginning of creation, therefore, if one adopts Ross's view, then the words of Jesus are incorrect, and He was wrong about the beginning of the world, humanity, and suffering. However, John stated that Jesus was present at the creation of the world. Thus, Jesus would have known precisely when He, the Father, and the Holy Spirit created the world. Even if one were to account for Jesus's humanity veiling some knowledge (e.g., the time of His return, Matt. 24:35–37), that veiling would be different than Him misleading others about the creation date of the world.

On the other hand, Jesus did not suggest that He created Adam and Eve as much as 135,000 years ago, nor did he suggest He created the universe billions of years ago. On the contrary, in Luke 11:50–51, Jesus used ἀπὸ καταβολῆς κόσμου (translated "since the foundation of the world"). According to this passage, the foundation of the world includes the creation of Adam and Eve, their disobedience, and the death of Abel.[372]

The genealogical records in Genesis give readers parameters for understanding the creation date. Though the Scriptures did not establish an exact date when God created Adam and Eve, the Scriptures revealed in Genesis 5 and 11 the genealogies from Adam to Abraham. The genealogical records of Genesis 5 and 11 equal about 2,000 years. Abraham was born around 2,000 BC, and about 2,000 years has elapsed since the birth of Jesus.[373] The age of the earth, then, is approximately 6,000 years old.[374] Thus, Jesus's words ἀπὸ καταβολῆς κόσμου harmonized more closely with the SDC view than with the PDAC view.

---

[372] I. Howard Marshall, *The Gospel of Luke: A Commentary on the Greek Text*, *The New International Greek Testament Commentary* (Exeter: Paternoster Press, 1978), 505.

[373] Based upon 1 Kings 6:1, this writer dates the birth of Abraham between 2,300 to 2,100 BC.

[374] Some proponents of SDC affirm an upper limit of the earth's age at 10,000. For an explanation of the non-chronogeological view, see David McGee, "Creation Date of Adam from the Perspective of Young-Earth Creationism," *Answers Research Journal* 5 (2012): 217–30.

If, for the sake of argument, Jesus believed in the PDAC view as implied by Ross, then this author would conclude that Jesus would have been a deceitfully misleading Messiah. He would have misinformed billions of people into following His leadership while secretly misleading people into thinking the earth was thousands of years old when He knew the earth was actually billions of years old. A disingenuous Jesus erodes His identity as God. Jesus cannot be trustworthy and intentionally deceive His followers about when He created Adam and Eve. Jesus could have chosen not to know something when He incarnated, but based upon His character, He cannot mislead. [375]

If one were to follow the old-earth creation model of Ross consistently, then one would have to conclude that Jesus misinformed His followers about the meaning of ἀπὸ καταβολῆς κόσμου. Jesus stated that ἀπὸ καταβολῆς κόσμου included the death of Abel. If Abel's death were part of the world's beginning, then reason would conclude that the creation of Adam and Eve was also part of the beginning of the world. However, following the view of Ross, the ἀπὸ καταβολῆς κόσμου was billions of years before the creation of Adam and Eve.[376] John's declaration that Jesus created the world would add more complexity and doubt to the trustworthy character of Christ. In other words, an obfuscated understanding of the creation event leads to doubt about Jesus's veracity. Ross has not recognized the eroding power of his view, if applied consistently, to the doctrine of Christology.

---

[375] Paul Copan, "Did God Become a Jew? A Defense of the Incarnation," in *Contending with Christianity's Critics*, ed. Paul Copan (Nashville: B&H, 2009), 227–30.

[376] Surprisingly, Ross did not comment about Jesus's words ἀπὸ καταβολῆς κόσμου. The author can only conclude that Ross did not see the incongruence between his view and Jesus's words regarding the beginning of creation. One explanation is that his presupposition only to interpret texts that affirm his PDAC view might have prohibited him from considering any other interpretation.

ANTHROPOLOGY: HISTORICITY OF ADAM

To Ross's credit, he affirmed the historicity of Adam.[377] He remarked, "We view all humans as the descendants of two historical persons, Adam and Eve, specially created by God and uniquely bearing his image."[378] Regarding this topic, he cited Genesis 3:20, Acts 17:26, and Romans 5:13–19. Ross added, "During six long eras, God systematically introduced new life-forms as changing conditions permitted or even required. During the seventh–the human era–God ceased from his work of creating new life-forms."[379]

As previously mentioned, Ross dated the creation of Adam between 6,000 to 135,000 years ago. His dating of the historical Adam required a deciphering ability from this author. For example, in 1994, he declared God created Adam and Eve 10,000 to 35,000 years ago with outside limits of 6,000 to 60,000 years ago. In 2006, Ross proclaimed God created them approximately 50,000 years ago. In 2014, Ross asserted that God created them 60,000 to 100,000 years ago. In 2017, he pronounced their creation date was between 12,000 to 135,000 years ago.[380]

Compounding Ross's ever-changing dating of the creation of Adam is Ross's inaccurate dating of Noah's flood at 40,000 years ago.[381] He affirmed the date of 40,000 years two times (2014 and 2017) in his writings.[382] Based on Ross's most recent writings, God created Adam and Eve 60,000 to 100,000 years ago.[383] According to Ross, if applied consistently, Adam's creation date cannot be earlier than 60,000 years ago.

Although Ross's view affirmed Adam's historicity, the Scriptures describe a different history than does Ross. First, according to the Scriptures, God did

---

[377] Ross, *Four Views*, 88.

[378] Ross, *Four Views*, 88.

[379] Ross, *Four Views*, 84.

[380] Ross, *Creation and Time*, 140; Ross, *Genesis One*, 21; Ross, *Navigating Genesis*, 75; Ross, *Four Views*, 92.

[381] Ross, *Navigating Genesis*, 156–57.

[382] Ross, *Four Views*, 85–86.

[383] Ross, *Navigating Genesis*, 75; Ross, *Four Views*, 85–86.

not create Adam 60,000 years ago. Intertextual commentary in Exodus 20:8–11, in Matthew 19:4, in Mark 10:6, and Acts 17:24–28 all affirm directly or indirectly that God created Adam no more than several thousand years ago, making the earth no older than 10,000 years.[384]

Second, according to Ross's writings, he has not been consistent with the creation date of Adam. Ross's followers cannot have an assurance that he will not change God's creation date of Adam as the supposed scientific evidence changes. Third, according to Ross, Adam's sin brought death upon humanity but not to the rest of creation. For Ross, death, disease, and bloodshed have existed since the inception of creation.[385] In Ross's view, the perfect conditions of Genesis 1:31 that God placed Adam and Eve in actually included billions of years of disease as well as animal, Neanderthal, and other hominid violence.[386]

According to Ross, humanity began at a different time and entered a world full of death and violence. The Scriptures do not affirm Ross's narrative of the creation of Adam. Although Ross has attempted to preserve Adam's historicity, Ross has adapted the biblical commentary into his PDAC view that is quite different from Genesis's description.

Thankfully, Ross has understood the theological significance of Adam. Adam's sin explained the need for Christ to redeem humanity. Nevertheless, there exists a sense in which the backdrop of Ross's Adam is not the same as the Adam of the Bible. Although a unique creation and distinct from all other creatures, Ross's Adam emerged when Neanderthals and hominids roamed the earth. According to Ross, humans, Neanderthals, and hominids coexisted.

---

[384] McGee, *Creation Date of Adam*, 228.

[385] Ross, *Four Views*, 86–87.

[386] Ross, *Four Views*, 88.

The Scriptures, however, do not assert that God created creatures similar to humans, such as Neanderthals and hominids.[387] Ross claimed that such creatures existed prior to the creation of Adam.[388] Ross's anthropology aligns with the Scriptures on the uniqueness of humanity originating from Adam and Eve. Ross did not argue that Adam evolved from a lower life form. He believes Adam is a special creation. Nevertheless, Ross's backdrop and timeline of humanity disagree with the Bible's backdrop and humanity's timeline.

SOTERIOLOGY: CONTEXT FOR THE GOSPEL

The setting of Ross's gospel message is quite different from what the authors of the Scriptures wrote:

> Physical death, though grievous, yields valuable redemptive benefits. Death of nonhuman life blessed humanity with a treasure chest of more than seventy-six quadrillion tons of biodeposits . . . from which to build a global civilization and facilitate the fulfillment of the Great Commission in mere thousands, rather than millions, of years. Christ's crucifixion and resurrection demonstrate, water baptism illustrates, and Paul repeatedly writes that only through death can we truly live, both now and forever.[389]

God states in Genesis 3, however, that death is a curse that is the result of Adam and Eve's sin. The world God pronounced very good was a world without death.

---

[387] This author struggles with Ross even introducing the concept of a hominid. Nowhere in the creation event does Moses imply a creature exists that is somewhere between an ape-like creature and a human.

[388] Ross, *Four Views*, 88.

[389] Ross, *Four Views*, 86–87.

Although Ross is correct that, through the death of Jesus, God blessed the world with the opportunity of forgiveness of sins and restoration with its Creator, the death of millions of animals and the decay of countless plants and vegetation was not a blessing. Death may relieve someone from pain and suffering, but death itself is not good. Ross appeared to deemphasize his world of billions of years of pain and suffering to justify death. According to Ross, plant and animal death is what he called a part of God's very good creation, providing coal, oil, and natural gas.[390] Ross employed the fallacy of equivocation with the word *death*.[391] He used *death* in two distinct manners: first, he used *death* to describe what happened voluntarily to Christ, a human. Second, he used *death* to describe what happens involuntarily to plants and animals.

Additionally, plants do not die in the same way animals die. When a landscaper cuts grass, people do not protest that thousands of blades of grass died. No mourning occurs after the landscaper discards the grass into a pile and it decomposes. A tree falls on the ground, and a year later, one might find individuals sitting on top of it to rest from a hike or sit and tie shoes. A year after a large elephant dies, however, one will not find people sitting on top of it to rest or to tie shoes because that would be disturbing to a human. Sitting on a dead tree is not. The decomposition of trees is not the same as the death of an animal. Chaffey and Lisle noted on the difference between animal and plant life:

> The Bible consistently makes a distinction between animal life and plants. Young-earth creationists point out that only creatures described as *nephesh chayyah* ("living soul" or "living creature") could not have died before the Fall. [Therefore], [s]ince plants are not alive

---

[390] Ross, *Four Views*, 86–87.

[391] Jason Lisle, *Introduction to Logic* (Green Forest, AR: Master Books), 78. The fallacy of equivocation is when one shifts from one meaning of a word to another within an argument. When applied to the Scriptures, a fallacy is a common error in reasoning that does not align with God's thinking.

in the biblical sense, then they cannot die in the biblical sense, either. Plants should be viewed as a sort of biological machine that God has created to sustain *nephesh* creatures on earth.[392]

The author of Genesis describes differently the constitution of animals and plants. Thus, Ross affirming the death of animals and plants in the same way is misleading to the reader of the Scriptures.

Ross never explains the origin of death, disease, and bloodshed for non-human life. Although, he did explain the horrific aspect of death for humanity and that Christ came to redeem humanity, he accepted that death and disease were part of God's very good creation that continues even to the present.[393] For Ross, death was an enemy for humanity while not necessarily the same enemy for nonhuman life.

The backdrop of Ross's gospel presentation was dissimilar to the historical presentation of the Bible. He claims, "Given that God intends to redeem billions of humans within just several thousands of years . . . Earth must be endowed with the resources not only to sustain billions of humans but also to support the technology required for Christ's followers to make disciples . . . from among all the world's people groups."[394] Ross argues that God had to allow the decay of plants and animals for billions of years to ensure the rapid transmission of the gospel and disciple-making since the industrial revolution because the period marked the point where humans needed biodeposits.

According to Ross, God did not flood the earth, which caused the rapid burial of plants and animals, and He was accountable to His laws of nature. Christ planned for animals to die for billions of years before He was ready to redeem people and, evidentially, the rest of creation. In Ross's gospel setting, he did not affirm Paul's words in Romans 8:19–23, where creation waits for

---

[392] Chaffey and Lisle, *Old-Earth Creationism*, 71.
[393] Ross, *Four Views*, 87.
[394] Ross, *Four Views*, 93.

God to release it from His curse because he did not assert that God cursed nonhuman life. Ross insisted that God would remove the curse, yet never explained the origin of the curse upon all the entire world. For Ross, death, disease, and bloodshed are part of God's original creation, which God declared very good.[395] Following Ross's reasoning, this writer does not comprehend the need to remove the curse at the end of time if/since it was very good at the beginning of time.

The doctrine of soteriology that Ross developed explains why God redeemed humanity but does not explain the need to redeem the rest of creation. Ross asserted that the death of animals and other nonhumans lasted until their corpses, coupled with decayed plants, produced enough biodeposits so that God could advance the spread of the gospel. The death of animals and decayed plants are the building blocks of global civilization that fulfill the Great Commission. God's best possible world, as Ross argued, includes billions of years of death and decay. Ross's soteriology has a point of contact with historic Christianity regarding humans, but his development of the doctrine of soteriology regarding the restoration of all of the world is dissimilar upon further investigation.[396]

ESCHATOLOGY: NEW HEAVEN AND EARTH WITHOUT A CURSE

Ross's eschatology harmonized with historic Christianity. He commented on the new heavens and earth: "Once the full number of humans comprising God's kingdom has been granted and received citizenship, evil and suffering will be permanently removed. [Then] God and his people and the angelic hosts will finally live together . . . death, decay, pain, and darkness (physical

---

[395] Ross, *Four Views*, 87.
[396] Ken Ham, *Six Days: The Age of the Earth and the Decline of the Church* (Green Forest, AR: Master Books, 2013), 200.

and spiritual) will be forever banished."[397] Based upon Revelation 21:1–5 and Isaiah 65:16–17, Ross presented an eschatology in which God will permanently remove the curse Moses described in Genesis.[398]

The new creation will no longer have pain, suffering, death, and violence associated with it. He noted, "God and his people and the angelic hosts will finally live together, face to face, experiencing unbroken fellowship and perfect love."[399] Ross interpreted John's words in Revelation 21:1–5 with a straightforward interpretation that reflected the author's intended meaning. His eschatology aligned with historic-orthodox Christianity.

Although Ross's protology did not align with the Scriptures, his eschatology did. Ross asserted that God would remove the curse upon all creation in the new earth. Ross never described the reason creation suffers, though he did assert that God will create a new heaven and earth void of pain, death, and sin. He based his view upon Revelation 21:1–5 and Isaiah 65:16–17. Ross aligned his eschatology with historic Christianity.

APOLOGETICS: A LACK OF CONSISTENCY

The subject of PDAC and SDC is not only biblical and theological but also an apologetic topic. William Craig defines apologetics as, "that branch of Christian theology which seeks to provide a rational justification from the Christian faith's truth claims. Apologetics is thus primarily a theoretical discipline, though it has a practical application."[400] Peter wrote that believers should always prepare themselves to give a defense for the reason of their hope of Christ (1 Pet. 3:15). Nancy Pearcy commented on 1 Peter 3:15, "The Greek word for 'defense' is apologia (the root word in apologetics) and it was originally a legal term, meaning the defendant's reply to the prosecutor in a

---

[397] Ross, *Four Views*, 87.

[398] Ross, *Four Views*, 87.

[399] Ross, *Four Views*, 87.

[400] William Lane Craig, *Reasonable Faith: Christian Truth and Apologetics*, 3rd ed. (Wheaton, IL: Crossway, 2008), 15.

court of law. Later the same term was used of the early Christian apologists . . . who defended the new faith against rampant paganism of the Roman Empire."[401] Among other topics, Christians must be ready to answer questions about the origins of the universe, the earth, and humanity.

Apologetically, Ross lacked consistency. He was not uniform with the application of special and general revelation and the field of hermeneutics. Ross also accused six-day creationists of promoting damaging theology within the body of Christ.

Moreover, Ross asserts that the Bible taught a "dual, reliably consistent revelation [and] one revelation of God's truth cannot be held as inferior or superior to another."[402] Adding to the lack of clarity, he argues, "According to both the Bible and the records of nature and history, God is responsible for the words of the Bible."[403] It appears Ross permits the observations of general revelation to contradict the interpretation of special revelation.

On the one hand, Ross asserted his belief in the magisterial role of special revelation. On the other hand, he claimed the equality, and at times, the magisterial role of general revelation. He claims that he relied upon scientific observations and the Bible to determine the earth's age. Yet, this book has revealed that exegetically the Bible does not permit a reader to interpret the earth as billions of years old. Ross's bibliology is not orthodox because Ross relegated the Bible to a ministerial role that is subject to scientists' ever-changing views.

Ross also lacks consistency in his method of interpretation. His protology depends upon his interpretation of observations from nature and the Bible, but his eschatology depends primarily upon his interpretation of the Scriptures. Ross asserts, "Neither here [Romans 5:12] nor anywhere else in the Scriptures does God's word say that Adam's offense brought death to all

---

[401] Nancy Pearcy, *Total Truth: Liberating Christianity from Its Cultural Captivity* (Wheaton, IL: Crossway, 2004), 124.

[402] Ross, *Creation and Time*, 56–57.

[403] Ross, *Creation and Time*, 58.

life (emphasis Ross)."[404] Ross employs a hermeneutic that fits within his presupposition that death, disease, and bloodshed did not begin when God cursed all of creation. Yet, Ross concludes that eschatologically God would remove the curse.[405] In other words, he was inconsistent with his method of interpretation. One moment he has followed the straightforward reading of the Scriptures (e.g., eschatology), and the next moment he has followed scientific philosophy by attempting to fit it into the Scriptures (e.g., protology).

Ross has developed an orthodox eschatology based solely upon the straightforward reading of the Bible. However, if one were to follow his protological hermeneutical method, any Christian doctrine might change that conflicts with future scientific observations. This writer would argue that Ross has already compromised on the straightforward reading of the creation event because of his presupposition that men's philosophical observations regarding the earth's age are correct.

Ross's most disturbing apologetic reflection regarding the age of the universe and the gospel is his view that the SDC view is analogous to the circumcision debate that the early church debated in Acts 15. Ross asserts, "As circumcision distorted the gospel and hampered evangelism, so, too, does young-universe creationism."[406] Ross has equated SDC to a belief rejected by the apostles at the Jerusalem Council. Paul has argued that circumcision distorts the gospel; Ross has argued that SDC similarly distorts the gospel.

Following the logic of Ross's words, he implies that the SDC view hinders the gospel's advancement. Thus, believers should accept the PDAC view, just like the early Christians accepted the Jerusalem Council's findings. To reject the SDC view, according to Ross's logic, is to reject a stumbling block that prohibits people from having a relationship with Christ.

---

[404] Ross, *Four Views*, 86.

[405] Ross, *Four Views*, 86.

[406] Ross, *Creation and Time*, 162.

The apostle Paul called people who caused stumbling in regard to the gospel as accursed. Ross's assertion implies that the PDAC view is orthodox, and the SDC view is unorthodox and accursed. The SDC view is the position that the church embraced until the eighteenth century. Ross's analogy would indicate that the differences between PDAC and SDC do not constitute a healthy debate but rather a theological war where only one side can be orthodox.[407]

Ross has elevated scientific observations and interpretations above the creation narrative as a decidedly unorthodox practice. The logical conclusion of Ross's apologetic method, if applied consistently, may cultivate errant doctrines that the reader would never be able to correct.

SUMMARY OF THE THEOLOGICAL IMPLICATIONS OF ROSS'S VIEW

The implications of Ross's PDAC view led to various affirmations and denials of orthodox Christianity. In the area of bibliology, Ross affirmed his interpretations of the creation event as the clear, straightforward, and literal reading of Genesis based upon the belief that scientific observations were more accurate than the SDC view. For Ross, the authority of the Scriptures in practice was often ministerial to the philosophical implications of scientists.

Ross's theology proper leads to the conclusion that the reader might not trust the words of the Scriptures. God inspired the author of Genesis to communicate that God created the earth in six literal days, but Ross revealed that God created and filled the earth over billions of years. Ross's interpretation of Genesis infers a God who wrote so poorly that readers might think His words communicated a recent creation when His words,

---

[407] David McGee, "Critical Analysis of Hugh Ross' Progressive Day-Age Creationism Through the Framework of Young-Earth Creationism," *Answers Research Journal* 12 (2019): 70–71.

according to Ross's view, actually communicated an ancient creation which is billions of years old.

Ross's Christology suggests that Jesus misdirected His followers about the beginning of creation. Jesus created the world and indicated that Adam, Eve, and Abel were part of the beginning years of creation at approximately six thousand years ago. According to Ross's view, the earth had existed for billions of years before humanity appeared.

This view would describe Jesus as one who misleads, thereby casting doubt on His deity. Following the implications of Ross's Christology, Jesus is a religious leader who has misinformed millions of people by means of His misrepresented claims.

The implications of Ross's anthropology are a mixture of orthodoxy and unorthodox affirmations. He affirmed the historicity of Adam and the need for Christ to redeem humanity. Nevertheless, after aligning with the Scriptures by insisting that Adam was a special creation of God, Ross deviated and declared that God created Adam between 40,000 to 135,000 years ago. According to Ross, death, disease, and bloodshed paralleled the inception of plant, animal, and hominoid life. Humankind experienced death and suffering after Adam's sin.

Ross's soteriology combines truth and fiction. The truth was humanity's sinfulness and need for a Savior. Christ came to the world because Adam's sin affected all of his descendants except Jesus by requiring the punishment of eternal separation from God. The fiction was Ross's emphasis upon the death and suffering of the rest of creation since the world began. The need for a redeemed creation was missing in Ross's view.

In the area of eschatology, Ross presents orthodox Christianity. He affirmed that no curse would exist in the new earth. The difficulty of harmonizing Ross's eschatology was that he never explained the origin of the curse and the need to remove it. According to Ross, God would create a new earth superior to the original one. This aspect of his eschatology developed from a literal reading of the Scriptures and aligned with historic-orthodox Christianity.

Apologetically, Ross lacks consistency. He has not been consistent with the application of special and general revelation and hermeneutics. Ross claims the Bible is authoritative over general revelation yet also claims that general revelation is like the Bible's sixty-seventh book. He affirms the equality of revelation from nature and the Bible but does not consistently apply his method to the creation narrative. In practice, his apologetic method gives deference to scientific philosophy over the Scriptures. Ross's most disturbing theological reflection regarding the universe's age was his assertion that the SDC view hinders the gospel's advancement. Inferred from Ross's assertation, believers should reject the SDC view, just as the first-century church rejected circumcision. Ross is correct about the vast differences and theological implications between the PDAC and SDC views. However, he is theologically incorrect in his assertion that the PDAC view is superior.

Bibliography

Books

Adams, Jay. *Is All Truth God's Truth?* Stanley, NC: Timeless Texts, 2003.

Akin, Daniel, and Paige Patterson. "The Doctrine of Christ." In *A Theology for the Church*, edited by Daniel L. Akin, 545–602. Nashville: B & H, 2007.

Allison, Gregg. *Historical Theology*. Grand Rapids: Zondervan, 2011.

St. Ambrose. "The First Day." In *Hexameron, Paradise, Cain and Abel*. Vol. 42, *The Fathers of the Church*, translated by John Savage. Washington, DC: Catholic University of America, 1961.

Archer, Gleason. *A Survey of Old Testament Introduction*. Rev. ed. Chicago: Moody, 2007.

Arnold, Bill T., and Bryan E. Beyer. *Encountering the Old Testament*. Grand Rapids: Baker, 2008.

Ashton, John F., ed. *In Six Days: Why Fifty Scientists Choose to Believe in Creation*. Green Forest, AR: Master Books, 2001.

Augustine of Hippo. "The Confessions of St. Augustine." In *Confessions and Letters of St. Augustin with a Sketch of His Life and Work*, edited by Philip Schaff, translated by J. G. Pilkington. Vol. 1 of *A Select Library of the Nicene and Post-Nicene Fathers of the Christian Church, First Series*. Buffalo, NY: Christian Literature, 1886.

Augustine of Hippo. *The City of God, Books VIII–XVI*. Edited by Hermigild Dressler, translated by Gerald G. Walsh and Grace Monahan. Vol. 14, *The Fathers of the Church*. Washington, DC: The Catholic University of America Press, 1952.

Barrick, William D. "Noah's Flood and Its Geological Implications." In *Coming to Grips with Genesis: Biblical Authority and the Age of the Earth,* edited by Terry Mortenson and Thane H. Ury, 251–82. Green Forest, AR: Master Books, 2008.

Barrick, William D. "Old Testament Evidence for a Literal, Historical Adam and Eve." In *Searching for Adam: Genesis & the Truth About Man's Origin*, edited by Terry Mortenson, 17-52. Greenfield, AR: Master Books, 2016.

Basil of Caesarea. "The Hexaemeron." In *St. Basil: Letters and Select Works*, edited by Philip Schaff and Henry Wace, translated by Blomfield Jackson. Vol. 8, *A Select Library of the Nicene and Post-Nicene Fathers of the Christian Church, Second Series*. New York: Christian Literature, 1895.

Beall, Todd. S. "Contemporary Hermeneutical Approaches to Genesis 1–11." In *Coming to Grips with Genesis: Biblical Authority and the Age of the Earth*, edited by Terry Mortenson and Thane H. Ury, 131–62. Green Forest, AR: Master Books, 2008.

Bediako, Daniel. *Genesis 1:1–2:3: A Textlinguistic Analysis*. Saarbrucken, Germany: VDM Verlag Dr. Müller, 2011.

Behe, Michael. *Darwin's Black Box*. New York: Free, 2006.

Bergman, Jerry. "Evolution, Racism, and the Scientific Vindication of Genesis." In *Searching for Adam: Genesis and the Truth About Man's Origin*, edited by Terry Mortenson, 375–414. Greenfield, AR: Master Books, 2016.

Boice, James Montgomery. *Foundations of the Christian Faith*. Downers Grove, IL: InterVarsity, 1986.

Bray, Gerald. *The Doctrine of God*. Downers Grove, IL: InterVarsity, 1993.

Brown, Fancis, S. R. Driver, and C. A. Briggs. *Brown-Driver-Briggs Hebrew and English Lexicon*. Peabody, MA: Hendrickson, 1996.

Bush, L. Russ, and Tom Nettles. *Baptists and the Bible*. Rev. ed. Nashville: B & H Academic, 1999.

Catchpoole, David, and Mark Harwood. "Ethics and Morality." In
    Evolution's Achilles' Heels, edited by Robert Carter, 233–60. Powder
    Springs, GA: Creation Book, 2014.

Chafer, Lewis Sperry. *Systematic Theology*. Vols. 1, 2, 5, and 6. Grand Rapids:
    Kregel, 1993.

Chaffey, Tim and Jason Lisle. *Old Earth Creationism on Trial: The Verdict Is In*.
    Green Forest, AR: Master Books, 2008.

Collins, C. J. *Did Adam and Eve Really Exist?* Wheaton, IL: Crossway Books,
    2011.

Comte de Buffon. *The Epochs of Nature*. Edited and translated by Jan
    Zalasiewicz, Anne-Sophie Milon, and Mateusz Zalasiewicz. Chicago:
    University of Chicago Press, 2018.

Copan, Paul. "Did God Become a Jew? A Defense of the Incarnation." In
    Contending with Christianity's Critics, edited by Paul Copan and William
    Lane Craig, 218–32. Nashville: B & H, 2009.

Coppes, Leonard. "852 יוֹם." In *Theological Wordbook of the Old Testament*,
    edited by R. Laird Harris, Gleason L. Archer Jr., and Bruce Waltke.
    Chicago: Moody, 1999.

Cowan, Steven B., and James S. Spiegel. *The Love of Wisdom*. Nashville: B
    & H, 2009.

Craig, William Lane. *Reasonable Faith: Christian Truth and Apologetics*. 3rd ed.
    Wheaton: Crossway Books, 2008.

Craig, William Lane. "God Directs All Things on Behalf of a Molinist View
    of Providence." In *Four Views on Divine Providence*, edited by Stanley N.
    Gundry and Dennis W. Jowers, 79–100. Grand Rapids: Zondervan,
    2011.

Darwin, Charles. *The Origin of Species*. New York: New American Library,
    2003.

Davidson, Richard. "The Genesis Account of Origins." In *The Genesis
    Creation Account: and Its Reverberations in the Old Testament*, edited by G.

A. Klingbeil, 59–130. Berrien Springs, MI: Andrews University Press, 2015.

Davis, J. J. *Paradise to Prison: Studies in Genesis*. Grand Rapids: Baker, 1984.

Davis, J. J. *Moses and the Gods of Egypt: Studies in Exodus*. 2nd ed. Winona Lake, IN: BMH Books, 1986.

DeYoung, Don. *Thousands . . . Not Billions: Challenging the Icon of Evolution, Questioning the Age of the Earth*. Green Forest, AR: Master Books, 2005.

Edgar, William, and K. Scott Oliphint, eds. *Christian Apologetics Past and Present: A Primary Source Reader, To 1500*. Vol. 1. Wheaton, IL: Crossway, 2009.

St. Ephrem the Syrian. *Selected Prose Works: Commentary on Genesis, Commentary on Exodus, Homily on Our Lord, Letter to Publius*, edited by Kathleen McVey, translated by Edward G. Matthews, Jr., Joseph P. Amar, Kathleen McVey. Vol. 91, *The Fathers of the Church*. Washington, DC: The Catholic University of America Press, 2010.

Erickson, Millard J. *Christian Theology*. 2nd ed. Grand Rapids: Baker Academics, 1998.

Estep, James R., Jr., Michael J. Anthony, and Gregg R. Allison. *A Theology for Christian Education*. Nashville: B & H, 2008.

Faber, George Stanley. *A Treatise on the Genius and Object of the Patriarchal, the Levitical, and Christian Dispensations*. Vol. 1. Whitefish, MT: Kessinger, 2009.

Fee, Gordon. D. and Douglas Stuart. *How to Read the Bible for All Its Worth*. 3rd ed. Grand Rapids: Zondervan, 2003.

Fields, Weston. *Unformed and Unfilled: A Critique of the Gap Theory*. Green Forest, AR: Master Books, 2005.

Frame, John M. *Apologetics: A Justification of Christian Belief*. Phillipsburg, NJ: P & R, 2015.

Freeman, Rick. "Do the Genesis Genealogies Contain Gaps?" In *Coming to Grips with Genesis: Biblical Authority and the Age of the Earth*, edited by

Terry Mortenson and Thane H. Ury, 283–314. Green Forest, AR: Master Books, 2008.

Garrett, James Leo, Jr. *Systematic Theology.* Vol. 2. 2nd ed. North Richland Hills, TX: BIBAL Press, 2001.

Garrett, James Leo, Jr. *Baptist Theology: A Four-Century Study.* Macon, GA: Mercer University, 2009.

Geisler, Norman L. *Systematic Theology.* Vols. 2–4. Minneapolis: Bethany House, 2004.

Geisler, Norman L. *Christian Apologetics.* Grand Rapids: Baker Academic, 2013.

Geisler, Norman. L. and W. C. Roach. *Defending Inerrancy: Affirming the Accuracy of Scripture for a New Generation.* Grand Rapids: Baker Books, 2011.

Gingrich, Roy E. *The Books of Haggai and Zechariah.* Memphis, TN: Riverside, 1999.

Greenwood, Kyle. *Scripture and Cosmology: Reading the Bible Between the Ancient World and Modern Science.* Downers Grove, IL: IVP Academic, 2015.

Gregory of Nyssa. *Hexaemeron. Patrologia Graeca.* Vol. 44. Edited by J. P. Migne. Translated by Richard McCambly. Paris: Migne, 1863.

Groothuis, Douglas. *Christian Apologetics.* Downers Grove, IL: InterVarsity, 2011.

Grudem, Wayne. *Systematic Theology.* Grand Rapids: Zondervan, 1994.

Gwynn, John. "Ephraim the Syrian and Aphrahat: Introductory Dissertation." In *Gregory the Great (Part II), Ephraim Syrus, Aphrahat,* edited by Philip Schaff and Henry Wace. Vol. 13, *A Select Library of the Nicene and Post-Nicene Fathers of the Christian Church, Second Series.* New York: Christian Literature Company, 1898.

Ham, Ken. *Six Days: The Age of the Earth and the Decline of the Church.* Green Forest, AR: Master Books, 2013.

Ham, Ken., Greg Hall, and Todd Hillard. *Already Compromised.* Green Forest, AR: Master Books, 2011.

Ham, Ken. "Young-Earth Creationism." In *Four Views on Creation, Evolution, and Intelligent Design*, edited by J. B. Stump, 17–48. Grand Rapids: Zondervan, 2017.

Ham, Ken, and Bodie Hodge. *A Flood of Evidence.* Green Forest, AR: Master Books, 2016.

Ham, Ken, Tim Lovett, Andrew A. Snelling, and John Whitmore. *A Pocket Guide to the Global Flood.* Hebron, KY: Answers in Genesis, 2009.

Hammet, John S. "Human Nature." In *A Theology for the Church*, edited by Daniel Akin, 340–408. Nashville, TN: B & H, 2007.

Hanna, William. *Memoirs of Thomas Chalmers.* Vol. 1. Scotland: Edmonston and Douglas, 1867.

Harbin, Michael. *The Promise and the Blessing: A Historical Survey of the Old and New Testaments.* Grand Rapids: Zondervan, 2005.

Hindson, Ed, and Gary Yates. *The Essence of the Old Testament: A Survey.* Nashville: B & H Academic, 2012.

Hirsch, E. D. *Validity in Interpretation.* New Haven, CT: Yale University Press, 1967.

Hodge, Bodie. "How Old is the Earth?" In *The New Answers Book 2*, edited by Ken Ham, 41–52. Green Forest, AR: Master Books, 2010.

Horton, Michael. *The Christian Faith.* Grand Rapids: Zondervan, 2011.

Irenaeus of Lyons. "Irenaeus against Heresies." In *Apostolic Fathers with Justin Martyr and Irenaeus*, edited by Alexander Roberts, James Donaldson, and A. Cleveland Coxe. Vol. 1, *Ante-Nicene Fathers.* Buffalo, NY: Christian Literature Company, 1885.

Ironside, H. A. *Studies on Book One of the Psalms.* Neptune, NJ: Loizeaux Brothers, 1952.

James, Garrett. *Inerrancy and the Southern Baptist Convention.* Dallas: Southern Baptist Heritage, 1986.

Johnson, Elliott E. *Expository Hermeneutics: An Introduction.* Grand Rapids: Zondervan, 1990.

Jones, F. N. *The Chronology of the Old Testament.* 15th ed. Green Forest, AR: Master Books, 2005.

Jordan, James B. *Creation in Six Days: A Defense of the Traditional Reading of Genesis One.* Moscow, ID: Canon, 1999.

Koehler, Ludwig, Walter Baumgartner, and Johann Jakob Stamm. *The Hebrew and Aramaic Lexicon of the Old Testament.* Leiden, Netherlands: E. J. Brill, 1994–2000.

Lactantius. "The Divine Institutes." In *Fathers of the Third and Fourth Centuries: Lactantius, Venantius, Asterius, Victorinus, Dionysius, Apostolic Teaching and Constitutions, Homily, and Liturgies*, edited by Alexander Roberts, James Donaldson, and A. Cleveland Coxe, translated by William Fletcher. Vol. 7, *Ante-Nicene Fathers.* Buffalo, NY: Christian Literature Company, 1886.

Leedy, P. D., and J. E. Ormrod. *Practical Research: Planning and Design.* 8th ed. Upper Saddle River, NJ: Prentice Hall, 2004.

Lewis, Gordon R. and Bruce A. Demarest. *Integrative Theology.* Vol. 1. Grand Rapids: Zondervan, 1987.

Lewis, Gordon R. and Bruce A. Demarest. *Integrative Theology.* Vol. 2. Grand Rapids: Zondervan, 1990.

Lightner, Robert. *Last Days Handbook.* Eugene, OR: Wipf and Stock, 2005.

Lisle, Jason. *Taking Back Astronomy: The Heavens Declare Creation and Science Confirms It.* Green Forest, AR: Master Books, 2012.

Lisle, Jason. *Introduction to Logic.* Green Forest, AR: Master Books.

Lisle, Jason. *The Physics of Einstein: Black holes, time travel, distant starlight, $E=mc^2$.* Aledo, TX: Biblical Science Institute, 2018.

Lyell, Charles. *Principles of Geology.* D. Appleton & Co: New York, 1830.

MacArthur, John. "Forward." In *Coming to Grips with Genesis: Biblical Authority and the Age of the Earth*, edited by Terry Mortenson and Thane H. Ury, 9–14. Green Forest, AR: Master Books, 2008.

Martyr, Justin. "Dialogue of Justin with Trypho, a Jew." In *Apostolic Fathers with Justin Martyr and Irenaeus*, edited by Alexander Roberts, James

Donaldson, and A. Cleveland Coxe. Vol. 1, *Ante-Nicene Fathers*. Buffalo, NY: Christian Literature Company, 1885.

Mason, Jim. "Radiometric Dating." In *Evolution's Achilles' Heels*, edited by Robert Carter, 193–214. Powder Springs, GA: Creation Book, 2014.

Mayr, Ernst. *What Evolution Is*. New York: Basic Books, 2001.

McCabe, Robert. V. "A Critique of the Framework Interpretation of the Creation Week." In *Coming to Grips with Genesis: Biblical Authority and the Age of the Earth*, edited by Terry Mortenson and Thane H. Ury, 211–50. Green Forest, AR: Master Books, 2008.

McDermott, Timothy S. "Aquinas, Thomas (1224/5–74)." *The Dictionary of Historical Theology*. Carlisle, Cumbria, UK: Paternoster, 2000.

McDowell, Josh, and Sean McDowell. *More Than a Carpenter*. Carol Stream, IL: Tyndale House, 2009.

Metzger, B. M. *The Text of the New Testament: Its transmission, Corruption, and Restoration*. 3rd ed. New York: Oxford University Press, 1992.

Meyer, Stephen C. *Darwin's Doubt*. Nashville: Harper Collins, 2013.

Mohler, R. Albert. "Confessional Evangelicalism." In *Four Views on the Spectrum of Evangelicalism*, edited by David Naselli and Collin Hansen, 68–96. Grand Rapids: Zondervan, 2011.

Mook, James. R. "The Church Fathers on Genesis, the Flood, and the Age of the Earth." In *Coming to Grips with Genesis: Biblical Authority and the Age of the Earth*, edited by Terry Mortenson and Thane H. Ury, 23–52. Green Forest, AR: Master Books, 2008.

Moreland, J. P. *Three Views on Creation and Evolution*. Grand Rapids: Zondervan, 1999.

Morris, Henry M. *The Genesis Record: A Scientific and Devotional Commentary on the Book of Beginnings*. Grand Rapids: Baker, 1976.

Morris, Henry M. *History of Modern Creationism*. 2nd ed. Santee, CA: Institute for Creation Research, 1993.

Morris, Henry M., and Gary E. Parker. *What is Creation Science?* Rev. ed. Green Forest, AR: Master Books, 1987.

Morris, Henry M. *The Biblical Basis for Modern Science.* Green Forest, AR: Master Books, 2002.

Mortenson, Terry. "'Deep Time' and the Church's Compromise: Historical Background." In *Coming to Grips with Genesis: Biblical Authority and the Age of the Earth*, edited by Terry Mortenson and Thane H. Ury, 79–104. Green Forest, AR: Master Books, 2008.

Mortenson, Terry. *The Great Turning Point.* Green Forest, AR: Master Books, 2004.

Mortenson, Terry, ed. *Searching for Adam: Genesis and the Truth About Man's Origin.* Green Forest, AR: Master Books, 2016.

Noll, Mark A. and David N. *Livingston, Evolution, Science, and Scripture: B. B. Warfield, Selected Writings.* Grand Rapids: Baker, 2000.

Origen. "Origen against Celsus." In *Fathers of the Third Century: Tertullian, Part Fourth; Minucius Felix; Commodian; Origen, Parts First and Second*, edited by Alexander Roberts, James Donaldson, and A. Cleveland Coxe, translated by Frederick Crombie. Vol. 4, *Ante-Nicene Fathers*. Buffalo, NY: Christian Literature Company, 1885.

Osborne, G. R. *The Hermeneutical Spiral: A Comprehensive Introduction to Biblical Interpretation.* Downers Grove, IL: InterVarsity, 1991.

Pearcy, Nancy. *Total Truth: Liberating Christianity from Its Cultural Captivity.* Wheaton, IL: Crossway, 2004.

Purtill, Richard L. "Defining Miracles." In *Defense of Miracles: A Comprehensive Case for God's Action in History*, edited by R. Douglas Geivett and Gary R. Habermas, 61–72. Downers Grove, IL: IVP Academic, 1997.

Rana, Fazale and Hugh Ross. *Origins of Life: Biblical and Evolutionary Models Face Off.* Colorado Springs: NavPress, 2004.

Robertson, A. T. *Word Pictures in the New Testament.* Nashville, TN: B & H, 1933.

Ross, Allen P. *Creation and Blessing: A Guide to the Study and Exposition of Genesis.* Grand Rapids: Baker Academic, 1997.

Ross, Hugh. *Creation and Time: A Biblical and Scientific Perspective on the Creation-Date Controversy*. Colorado Springs: NavPress, 1994.

Ross, Hugh. *Genesis One: A Scientific Perspective*. 4th ed. Covina, CA: Reasons to Believe, 2006.

Ross, Hugh. *Navigating Genesis: A Scientist's Journey through Genesis 1–11*. Covina, CA: RTB Press, 2014.

Ross, Hugh. "Old-Earth (Progressive) Creationism." In *Four Views on Creation, Evolution, and Intelligent Design*, edited by J. B. Stump, 71–100. Grand Rapids: Zondervan, 2017.

Ross, Marcos. *The Heavens and the Earth: Excursions in Earth and Space Science*. Dubuque, IA: Kendal Hunt, 2015.

Ryrie, Charles. *Basic Theology*. Wheaton: Victor Books, 1986.

Sailhamer, John. H. *The Meaning of the Pentateuch: Revelation, Composition and Interpretation*. Downers Grove, IL: InterVarsity, 2009.

Sarfati, Jonathan. *Refuting Compromise*. Green Forest, AR: Master Books, 2004.

Scofield, C. I. *Scofield Study Bible*. New York: Oxford University Press, 1909.

Smith, David L. *A Handbook of Contemporary Theology*. Grand Rapids: Victor Books, 1992.

Snelling, Andrew. *Earth's Catastrophic Past: Geology, Creation, and the Flood*. Vol. 1. Dallas: Institute for Creation Research, 2009.

Snelling, Andrew. "10 Best Evidences from Science that Confirm a Young Earth." In *Best Evidences: A Pocket Guide*. Hebron, KY: Answers in Genesis, 2013.

Sproul, R. C. *The Consequences of Ideas*. Wheaton, IL: Crossway Books, 2000.

Sproul. R. C. *Defending Your Faith: Introduction to Apologetics*. Wheaton, IL: Crossway, 2003.

Sproul, R. C. *Explaining Inerrancy*. Orlando, FL: Ligonier Ministries, 1996.

Strobel, Lee. *The Case for Christ*. Grand Rapids: Zondervan, 1998.

Stump, J. B., ed. *Four Views on Creation, Evolution, and Intelligent Design*. Grand Rapids: Zondervan, 2017.

Ussher, James. *The Annals of the World.* 1658. Reprint. Green Forest, AR: Master Books, 2006.

Victorinus of Pettau. "On the Creation of the World." In *Fathers of the Third and Fourth Centuries: Lactantius, Venantius, Asterius, Victorinus, Dionysius, Apostolic Teaching and Constitutions, Homily, and Liturgies,* edited by Alexander Roberts, James Donaldson, and A. Cleveland Coxe, translated by Robert Ernest Wallis. Vol. 7, *Ante-Nicene Fathers.* Buffalo, NY: Christian Literature Company, 1886.

Walton, John H. *The Lost World of Genesis One: Ancient Cosmology and the Origins Debate.* Downers Grove, IL: IVP Academic, 2009.

Whitcomb, John. C. *The Early Earth: An Introduction to Biblical Creationism.* 2nd ed. Rev. ed. Winona Lake, IN: BMH Books, 1986.

Whitcomb, John. C. and Henry M. Morris. *The Genesis Flood: The Biblical Record and Its Scientific Implications.* Phillipsburg, NJ: Presbyterian and Reformed, 1961.

Wise, Kurt. *Faith, Form, and Time.* Nashville: B & H, 2002.

Yeago, David S. "Luther, Martin (1483–1546)." *The Dictionary of Historical Theology.* Carlisle, Cumbria, UK: Paternoster, 2000.

Commentaries

Alden, Robert L. *Job.* Vol. 11 of *The New American Commentary,* edited by E. Ray Clendenen. Nashville: B & H, 1993.

Batten, Loring W. *A Critical and Exegetical Commentary on the Books of Ezra and Nehemiah.* In *International Critical Commentary,* edited by J. A. Emerton. New York: Scribner, 1913.

Beale, G. K. *The Book of Revelation: A Commentary on the Greek Text. The New International Greek Testament Commentary.* Grand Rapids: Eerdmans, 1999.

Best, Ernest. *A Critical and Exegetical Commentary on Ephesians.* In *International Critical Commentary,* edited by J. A. Emerton. Edinburgh: T & T Clark International, 1998.

Boa, Kenneth and William Kruidenier. *Romans*. Vol. 6 of *Holman New Testament Commentary*, edited by Max Anders. Nashville, TN: B & H, 2000.

Briggs, Charles A., and Emilie Grace Briggs. *A Critical and Exegetical Commentary on the Book of Psalms*. In *International Critical Commentary*, edited by J. A. Emerton. New York: C. Scribner's Sons, 1907.

Bruce, F. F. *Romans: An Introduction and Commentary*. Vol. 6 of *Tyndale New Testament Commentaries*, edited by Leon Morris. Downers Grove, IL: InterVarsity, 1985.

Buzzell, Sid S. "Proverbs." Vol. 1 of *The Bible Knowledge Commentary: An Exposition of the Scriptures*, edited by J. F. Walvoord and R. B. Zuck, 901–74. Wheaton, IL: Victor Books, 1985.

Calvin John. *Commentary on Genesis*. Vol. 1. Translated by John King. Grand Rapids: Christian Classic Ethereal Library, 1847.

Calvin, John, and Henry Beveridge. *Institutes of the Christian Religion*. Vol. 1. Edinburgh: The Calvin Translation Society, 1845.

Carpenter, Eugene. **Exodus**. Vol. 2 of *Evangelical Exegetical Commentary*, edited by H. Wayne House and William D. Barrick. Bellingham, WA: Lexham, 2012.

Carson, D. A. *The Gospel According to John*. In *The Pillar New Testament Commentary*. Grand Rapids: InterVarsity, 1991.

Christensen, Duane L. *Deuteronomy 1–21:9*. Rev. ed. Vol. 6A of *Word Biblical Commentary*, edited by Bruce M. Metzger. Dallas: Thomas Nelson, 2001.

Clendenen, E. Ray "The Minor Prophets." In *Holman Concise Bible Commentary*, edited by David S. Dockery. Nashville, TN: B & H, 1998.

Cole, R. Alan. *Exodus: An Introduction and Commentary*. Vol. 2 of *Tyndale Old Testament Commentaries*, edited by Leon Morris. Downers Grove, IL: InterVarsity, 1973.

Chisholm, Robert., Jr. "Hosea." Vol. 1 of *The Bible Knowledge Commentary: An Exposition of the Scriptures*, edited by J. F. Walvoord and R. B. Zuck, 1377–1408. Wheaton, IL: Victor Books, 1985.

Davids, Peter H. *The Letters of 2 Peter and Jude*. In *The Pillar New Testament Commentary*, edited by D. A. Carson. Grand Rapids: Eerdmans, 2006.

Deere, Jack S. "Deuteronomy." Vol. 1 of *The Bible Knowledge Commentary: An Exposition of the Scriptures*, edited by J. F. Walvoord and R. B. Zuck, 259–324. Wheaton, IL: Victor Books, 1985.

Driver, S. R. *A Critical and Exegetical Commentary on Deuteronomy*. 3rd ed. In *International Critical Commentary*, edited by J. A. Emerton. Edinburgh: T & T Clark, 1902.

Durham, John I. *Exodus*. Vol. 3 of *Word Biblical Commentary*, edited by Bruce M. Metzger. Dallas: Word, 1987.

Gangel, Kenneth O. *Acts*. Vol. 5 of *Holman New Testament Commentary*, edited by Max Anders. Nashville, TN: Broadman & Holman, 1998.

Gangel, Kenneth O. "2 Peter." Vol. 2 of *The Bible Knowledge Commentary: An Exposition of the Scriptures*, edited by J. F. Walvoord and R. B. Zuck, 859–80. Wheaton, IL: Victor Books, 1985.

Grassmick, John D. "Mark." Vol. 2 of *The Bible Knowledge Commentary: An Exposition of the Scriptures*, edited by J. F. Walvoord and R. B. Zuck, 95–198. Wheaton, IL: Victor Books, 1985.

Gray, George Buchanan. *A Critical and Exegetical Commentary on the Book of Isaiah, I–XXXIX*. In *International Critical Commentary*, edited by J. A. Emerton. New York: C. Scribner's Sons, 1912.

Green, Michael. *2 Peter and Jude: An Introduction and Commentary*. Vol. 18 of *Tyndale New Testament Commentaries*, edited by Leon Morris. Downers Grove, IL: InterVarsity, 1987.

Guthrie, Donald. *Hebrews: An Introduction and Commentary*. Vol. 15 of *Tyndale New Testament Commentaries*, edited by Leon Morris. Downers Grove, IL: InterVarsity, 1983.

Hagner, Donald A. *Matthew 14–28*. Vol. 33B of *Word Biblical Commentary*, edited by Bruce M. Metzger. Dallas: Word, Incorporated, 1995.

Hendriksen, William and Simon J. Kistemaker. *Exposition of Colossians and Philemon*. Vol. 6, *New Testament Commentary*, edited by William Hendriksen. Grand Rapids: Baker, 2001.

Hendriksen, William and Simon J. Kistemaker. *Exposition of Paul's Epistle to the Romans*. Vols. 12–13, of *New Testament Commentary*, edited by William Hendriksen. Grand Rapids: Baker, 2001.

Hicks, John Mark. *1 & 2 Chronicles*. In *College Press NIV Commentary*, edited by Terry Briley. Joplin, MO: College Press Pub. Co., 2001.

Hodges, Zane. "Hebrews." Vol. 2 of *The Bible Knowledge Commentary: An Exposition of the Scriptures*, edited by J. F. Walvoord and R. B. Zuck, 777–814. Wheaton, IL: Victor Books, 1985.

Hubbard, David A. *Proverbs*. Vol. 15 of *Preacher's Commentary Series*, edited by Lloyd J. Ogilvie. Nashville, TN: Thomas Nelson, 1989.

Ironside, H. A. *Studies on Book One of the Psalms*. Neptune, NJ: Loizeaux Brothers, 1952.

Keil, Carl Friedrich and Franz Delitzsch. *Commentary on the Old Testament*. Vol. 4. Peabody, MA: Hendrickson, 1996.

Kidner, F. Derek. "Isaiah." In *New Bible Commentary: 21st Century Edition*, edited by D. A. Carson. 4th ed. Downers Grove, IL: InterVarsity, 1994.

Kidner, Derek. *Psalms 1–72: An Introduction and Commentary*. Vol. 15 of *Tyndale Old Testament Commentaries*, edited by Leon Morris. Downers Grove, IL: InterVarsity, 1973.

Kistemaker, Simon J. and William Hendriksen. *Exposition of the Acts of the Apostles*. Vol. 17, *New Testament Commentary*. Grand Rapids: Baker, 2001.

Kistemaker Simon J., and William Hendriksen. *Exposition of Hebrews*. Vol. 15, *New Testament Commentary*. Grand Rapids: Baker, 2001.

Köstenberger, Andreas J. *John*. In *Baker Exegetical Commentary on the New Testament*. Grand Rapids: Baker, 2004.

Lactantius. "The Divine Institutes." In *Fathers of the Third and Fourth Centuries: Lactantius, Venantius, Asterius, Victorinus, Dionysius, Apostolic Teaching and Constitutions, Homily, and Liturgies*, edited by Alexander Roberts, James Donaldson, and A. Cleveland Coxe and translated by William Fletcher. Vol. 7, *Ante-Nicene Fathers*. Buffalo, NY: Christian Literature Company, 1886.

Lange, John Peter. *A Commentary on the Holy Scriptures: Hosea*. Bellingham, WA: Logos Bible Software, 1899, 2008.

Lowery, David K. "1 Corinthians." Vol. 2, *The Bible Knowledge Commentary: An Exposition of the Scriptures*, edited by J. F. Walvoord and R. B. Zuck, 505–50. Wheaton, IL: Victor Books, 1985.

Mangum, Douglas, ed. *Lexham Context Commentary: New Testament*. Bellingham, WA: Lexham, 2020.

Marshall, I. Howard. *The Gospel of Luke: A Commentary on the Greek Text*. In *The New International Greek Testament Commentary*. Exeter: Paternoster, 1978.

Martin, John A. "Isaiah." Vol. 1 of *The Bible Knowledge Commentary: An Exposition of the Scriptures*, edited by J. F. Walvoord and R. B. Zuck, 1029–1122. Wheaton, IL: Victor Books, 1985.

Mathews, K. A. *Genesis 1–11:26*. Vol. 1A of *The New American Commentary*, edited by E. Ray Clendene. Nashville: B & H, 1996.

Melick, Richard R. *Philippians, Colossians, Philemon*. Vol. 32 of *The New American Commentary*, edited by E. Ray Clendene. Nashville: B & H, 1991.

Merrill, Eugene H. "1 Chronicles." Vol. 1 of *The Bible Knowledge Commentary: An Exposition of the Scriptures*, edited by J. F. Walvoord and R. B. Zuck, 589–618. Wheaton, IL: Victor Books, 1985.

Morris, Leon. *The Epistle to the Romans*. In *The Pillar New Testament Commentary*, edited by D. A. Carson. Grand Rapids: Eerdmans, 2006.

Morris, Leon. *Revelation: An Introduction and Commentary.* Vol. 20 of *Tyndale New Testament Commentaries*, edited by Leon Morris. Downers Grove, IL: InterVarsity, 1987.

Motyer, J. Alec. *Isaiah: An Introduction and Commentary.* Vol. 20 of *Tyndale Old Testament Commentaries*, edited by Leon Morris. Downers Grove, IL: InterVarsity, 1999.

Nolland, John. *The Gospel of Matthew: A Commentary on the Greek Text. The New International Greek Testament Commentary*, edited by I. Howard Marshall and Donald A. Hagner. Grand Rapids: Eerdmans, 2005.

Osborne, Grant R. *Revelation. Baker Exegetical Commentary on the New Testament*, edited by Moises Silva. Grand Rapids: Baker, 2002.

Ross, Allen P. "Psalms." Vol. 1 of *The Bible Knowledge Commentary: An Exposition of the Scriptures*, edited by J. F. Walvoord and R. B. Zuck, 779–900. Wheaton, IL: Victor Books, 1985.

Sanday, W. and Arthur C. Headlam. *A Critical and Exegetical Commentary on the Epistle of the Romans.* 3rd ed. In *The International Critical Commentary*, edited by J. A. Emerton. New York: C. Scribner's Sons, 1897.

Schreiner, Thomas R. *Romans.* Vol. 6 of *Baker Exegetical Commentary on the New Testament*, edited by Moises Silva. Grand Rapids: Baker Books, 1998.

Skinner, John. *A Critical and Exegetical Commentary on Genesis.* In *The International Critical Commentary*, edited by J. A. Emerton. New York: Scribner, 1910.

Skinner, R. David. *Studies in Genesis 1–11: A Creation Commentary.* Edited by Michael R. Spradlin. Collierville, TN: Innovo, 2018.

Spence-Jones, H. D. M., ed. *Jeremiah.* Vol. 2, *Pulpit Commentary*. New York: Funk & Wagnalls, 1909.

Stuart, Douglas K. *Exodus.* Vol. 2 of *The New American Commentary*, edited by E. Ray Clendene. Nashville: Broadman & Holman, 2006.

Stuhlmueller, Carroll. *Rebuilding with Hope: A Commentary on the Books of Haggai and Zechariah. International Theological Commentary*. Grand Rapids: Eerdmans, 1988.

Thiselton, Anthony C. *The First Epistle to the Corinthians: A Commentary on the Greek Text. The New International Greek Testament Commentary*. Grand Rapids: Eerdmans, 2000.

Walton, John H. *Genesis*. In *The NIV Application Commentary*, edited by Terry Muck. Grand Rapids: Zondervan, 2001.

Walvoord, John F. "Revelation." Vol. 2 of *The Bible Knowledge Commentary: An Exposition of the Scriptures*, edited by J. F. Walvoord and R. B. Zuck, 925–91. Wheaton, IL: Victor Books, 1985.

Wenham, Gordon J. "Genesis." *New Bible Commentary: 21st Century Edition*, 4th ed., edited by D. A. Carson et al., 54–92. Downers Grove, IL: InterVarsity, 1994.

Wenham, Gordon J. *Genesis 1–15*. Vol. 1 of *Word Biblical Commentary*, edited by Bruce M. Metzger. Dallas: Word, 1987.

Wiersbe, Warren W. *With the Word Bible Commentary*. Nashville: Thomas Nelson, 1991.

Witmer, John A. "Romans." Vol. 2 of *The Bible Knowledge Commentary: An Exposition of the Scriptures*, edited by J. F. Walvoord and R. B. Zuck, 435–504. Wheaton, IL: Victor Books, 1985.

Wolff, Hans Walter. *Hermeneia: Hosea: A Commentary on the Book of the Prophet Hosea*. Translated by Gary Stansell. Philadelphia: Fortress, 1974.

Young, Edward. *The Book of Isaiah, Chapters 40–66*. Vol. 3. Grand Rapids: Eerdmans, 1972.

Zuck, Roy B. "Job." Vol. 1 of *The Bible Knowledge Commentary: An Exposition of the Scriptures*, edited by J. F. Walvoord and R. B. Zuck, 715–78. Wheaton, IL: Victor Books, 1985.

Journals

Arp, William. "Authorial Intent." *Journal of Ministry and Theology* 4, no. 1 (2002): 36–50.

Beall, Todd S. "Genesis 1–11: A Plea for Hermeneutical Consistency." *Bible and Spade* 29, no. 2 (2016): 68–74.

Bloomberg, Craig L. "Marriage, Divorce, Remarriage, And Celibacy: An Exegesis of Matthew 19:3–12." *Trinity Journal* 11, no 2 (1990): 161–96.

Davis, J. "24 Hours–Plain as Day." *Answers* 7, no. 2 (2012): 67–69.

Faulkner, Danny. "A Proposal for a New Solution to the Light Travel Time Problem." *Answers Research Journal* 6 (2013): 279–84.

Faulkner, Danny. "Thoughts on the rāqîa and a Possible Explanation for the Cosmic Microwave Background." *Answers Research Journal* 9 (2016): 57–65.

Faulkner, Danny. "Solving the Light Travel Time Problem," *Answers in Depth* 16 (2021).

Hartnett, John G. "Speculation of Redshift in a Created Universe." *Answers Research Journal* 8 (2015): 77–83.

Humphreys, D. Russell. "New Time Dilation Helps Creation Cosmology." *Journal of Creation* 22, no. 3 (2008): 84–92.

Lisle, Jason. "Anisotropic Synchrony Convention–A Solution to the Distant Starlight Problem." *Answers Research Journal* 3 (2010): 191–207.

MacLeod, David J. "The Adoration of God the Creator: An Exposition of Revelation 4." *Bibliotheca Sacra* 134, no. 654 (2007): 198–219.

Mcaffee, Matthew. "Creation and the Role of Wisdom in Proverbs 8: What Can We Learn?" *Southeastern Theological Review* 10, no. 2 (2019): 31–57.

McGee, David. "Critical Analysis of Hugh Ross' Progressive Day-Age Creationism Through the Framework of Young-Earth Creationism." *Answers Research Journal* 12 (2019): 53–71.

McGee, David. "Old Earth Theology: A Factor that Explains Inconsistent Belief of Inerrancy Among Florida Southern Baptists." *Answers Research Journal* 7 (2014): 363–86.

McGee, David. "Creation Date of Adam from the Perspective of Young-Earth Creationism." *Answers Research Journal* 5 (2012): 217–30.

Mortenson, Terry. "Jesus, Evangelical Scholars, and the Age of the Earth." *The Master's Seminary Journal* 18, no. 1 (2007): 69–99.

Mortenson, Terry. "Systematic Theology Texts and the Age of the Earth: A Response to the Views of Erickson, Grudem, and Lewis and Demarest." *Answers Research Journal* 2 (2009): 175–200.

Mortenson, Terry. "Inerrancy and Biblical Authority: How and Why Old-Earth Inerrantists Are Unintentionally Undermining Inerrancy." *Answers Research Journal* 13 (2012): 189–219.

Newton, Robert. "Distant Starlight and Genesis: Conventions of Time Measurement." *Journal of Creation* 15, no.1 (2001): 80–85.

Sarfati, J. "What about Cainan?" *Journal of Creation* 18, no. 2 (2004): 41–43.

Sexton, Jeremy. "Evangelicalism's Search for Chronological Gaps in Genesis 5 and 11: A Historical, Hermeneutical, and Linguistic Critique." *Journal of the Evangelical Society* 61, no. 1 (2018): 5–25.

Stallard, Michael. "Literal Interpretation: The Key to Understanding the Bible." *The Journal of Ministry and Theology* 4, no. 1 (2000): 14–35.

Vardiman, Larry, and D. Russell. Humphreys. "A New Creationist Cosmology: In No Time at All Part 1." *Acts & Facts* 39, no. 11 (2010): 12–15.

Vardiman, Larry, and D. Russell Humphreys. "A New Creationist Cosmology: In No Time at All Part 2." *Acts & Facts* 40, no. 1 (2011): 12–14.

Vardiman, Larry, and D. Russell Humphreys. "A New Creationist Cosmology: In No Time at All Part 3." *Acts & Facts* 40, no. 2 (2011): 12–14.

Waltke, Bruce. "The Creation Account in Genesis 1:1–3: Part II: The Restitution Theory." *Bibliotheca Sacra* 132, no. 526 (1975): 136–44.

Wardlaw, Terrance R., Jr. "The Significance of Creation in the Book of Isaiah." *Journal of the Evangelical Theological Society* 59, no. 3 (2019): 449–71.

Warfield, B. B. "On the Antiquity and the Unity of the Human Race." *The Princeton Theological Review* 9, no. 1 (1911): 1–25.

Whitcomb, John. "The Science of Historical Geology." *Westminster Theological Journal* 36 (1973): 65–77.

Electronic Documents

Armitage, Mark and Jim Solliday. "UV Autofluorescence Microscopy of Dinosaur Bone Reveals Encapsulation of Blood Clots within Vessel Canals." *Microscopy Today*, 28, no 5 (2020): 30–38. https://www.cambridge.org/core/ journals/microscopy-today/article/uv-autofluorescence-microscopy-of-dinosaur-bone-reveals-encapsulation-of-blood-clots-within-vessel-canals/8762E671960898DAC303973A5A2A93F6#.

Aquinas, Thomas. "First Part, Questions 69, 74." In *Summa Theologica*, translated by Fathers of the English Dominican Province, 772–77, 797–807. New York: Benziger Bros., 1947. https://www.ccel.org.

Aquinas, Thomas. "First Part, Questions 69." In *Summa Theologica*, translated by Fathers of the English Dominican Province, 772–77. New York: Benziger Bros., 1947. https://www.ccel.org.

Boyd, Steven. "Statistical Determination of Genre in Biblical Hebrew: Evidence for an Historical Reading of Genesis 1:1–2:3." In *Radioisotopes and the Age of the Earth: Results of a Young-Earth Creationist Research Initiative*, edited by Larry Vardiman, Andrew A. Snelling, and Eugene F. Chaffin, 631–732. Chino Valley, AZ: Creation Research Society. https://www.icr.org/article/statistical-determination-genre-biblical.

Comte de Buffon. "From the System of Whiston." Vol. 1, *Buffon's Natural History*, translated by James Smith Barr, 115. London: Paternoster-

Row, 2014. https://www.gutenberg.org/files/44792/44792-h/44792-h.htm.

Criswell, W. A. ed. *Believer's Study Bible*, electronic ed. Nashville: Thomas Nelson, 1991.

Fields, Helen, "Dinosaur Shocker." *Smithsonian Magazine*, April 30, 2006. https://www.smithsonianmag.com/science-nature/dinosaur-shocker-115306469/.

Freeman, Rick. "Do the Genesis Genealogies Contain Gaps?" Answers in Genesis. https://assets.answersingenesis.org/doc/articles/aid/v2/do-the-genesis-genealogies-contain-gaps.pdf.

Geisler, Norman L. "Does Believing in Inerrancy Require One to Believe in Young Earth Creationism?" Norman Geisler. http://normangeisler.com/does-believing-in-inerrancy-require-one-to-believe-in-young-earth-creationism/.

Hutton, James. "Chapter 1." Vol. 1, *Theory of the Earth*. Royal Society of Edinburgh: Scotland, 1788. https://www.gutenberg.org/files/12861/12861-h/12861-h.htm.

Ice, Thomas. D., and J. J. S. Johnson. "Using Scriptural Data to Calculate a Range-Qualified Chronology from Adam to Abraham." Paper presented at the Southwest Regional Meeting of the Evangelical Theological Society at Criswell College, Dallas, TX, March 1, 2002. http://www.icr.org/article/4639/.

Josephus, F. "Antiquities of the Jews—Book 1." In the *Works of Flavius Josephus*. Translated by W. Whiston. London: Ward, Lock & Bowden, 1897. http://www.ccel.org/ j/josephus/works/ant-1.htm.

Lamarck, Jean. *Zoological Philosophy*. Translated by Hugh Elliot. New York: Hafner, 1964. http://www.blc.arizona.edu/courses/schaffer/449/Lamarck/Lamarck%20Zoological%20Philosophy.pdf.

Luther, Martin. "Introduction." In *Commentary on Genesis*, edited by John Nicholas Lenker, translated by Henry Cole. Lutherans In All Lands: Minneapolis: MN, 1904.
http://www.gutenberg.org/files/48193/48193-h/48193-h.htm.

Luther, Martin. "Part IV. The Creation of Eve." In *Commentary on Genesis*, edited by John Nicholas Lenker, translated by Henry Cole. Minneapolis, MN: Lutherans In All Lands, 1904.
http://www.gutenberg.org/files/48193/48193-h/48193-h.htm.

Luther, Martin. "Part VI. God's Work on the Sixth Day." In *Commentary on Genesis*, edited by John Nicholas Lenker, translated by Henry Cole. Minneapolis, MN: Lutherans In All Lands, 1904.
http://www.gutenberg.org/files/48193/48193-h/48193-h.htmRana, Fazale. Who was Adam? An Old-Earth Creation Model for the Origin of Humanity. http://www.reasons.org/articles/who-was-adam-an-old-earth-creation-model-for-the-origin-of-humanity.

Report of the SBC Peace Committee. Report of the Southern Baptist Convention Peace Committee. In SBC Annual, item 153, St. Louis, MO: SBC, 1987. https:// www.baptist2baptist.net/ b2barticle.asp?id=65.

Schaff, Philip. Vol. 6, *Ante-Nicene Fathers*. Edinburgh, United Kingdom: T & T Clark, 1885. http://www.ccel.org/ccel/schaff/ anf06.v.v.vi.html.

Snelling, Andrew. "Excess Argo: The 'Achilles Heel' of Potassium-Argon and Argo-Argon Dating of Volcanic Rocks." *Acts & Facts* 28, No.1, 1999. https://www.icr.org/ article/excess-argon-achilles-heel-potassium-argon-dating.

Snelling, Andrew. "Global Evidences of the Genesis Flood." *Answers Magazine*. July 1, 2021. https://answersingenesis.org/the-flood/global/evidences-genesis-flood/.

Spurgeon, Charles. "Christ the Destroyer." Sermon, December 17, 1876. https://www.spurgeongems.org/ sermon/chs1329.pdf.

Spurgeon, Charles. "Unconditional Election." Sermon, September 2, 1855. https://www.spurgeongems.org/sermon/chs41-42.pdf.

Tomkins Jeffery P. "New Testament Upholds Created Kind Stasis." *Acts and Facts*, 49:10 (2020). https://www.icr.org/article/new-testament-upholds-created-kind-stasis/.

Wooddell, J. D. *The Baptist Faith and Message 2000: Critical Issues in America's Largest Protestant Denomination.* Edited D. K. Blount and J. D. Wooddell. Nashville, TN: B & H 2007. Kindle.

Unpublished Dissertations

McGee, David A. "A Mixed-Methods Study of the Variables that Influenced Florida Southern Baptist' Affirmation of the Inerrancy of the Bible." EdD diss., Southeastern Baptist Theological Seminary, 2014. ProQuest Dissertations & Theses Global.

Tison, Richard. "Lords of Creation: American Scriptural Geology and the Lord Brothers' Assault on 'Intellectual Atheism.'" PhD diss., University of Oklahoma, 2008. ProQuest Dissertations & Theses Global.

Williams, Randall. "The Role of the Peace Committee in the Southern Baptist Convention Inerrancy Controversy." PhD diss., Mid-America Baptist Theological Seminary, 2000. ProQuest Dissertations & Theses Global.

Reference Works

Bray, Gerald. "Clement of Alexandria (c. 150–c. 215)." *The Dictionary of Historical Theology.* Carlisle, Cumbria, UK: Paternoster, 2000.

Cross, F. L., and Elizabeth A. Livingstone, eds. *The Oxford Dictionary of the Christian Church.* New York: Oxford University Press, 2005.

Hogg, David S. "Anselm of Canterbury (1033–1109)." *The Dictionary of Historical Theology.* Carlisle, Cumbria, UK: Paternoster, 2000.

Ocker, Christopher. "Lactantius (c. 250–c. 324)." *The Dictionary of Historical Theology*. Carlisle, Cumbria, UK: Paternoster, 2000.

Swanson, James A. *Dictionary of Biblical Languages with Semantic Domains: Hebrew*. Oak Harbor, WA: Logos Research Systems, 1997.

Walker, James B. "Origen (c. 185–c. 254)." *The Dictionary of Historical Theology*. Carlisle, Cumbria, UK: Paternoster, 2000.

Wilson, Henry Austin. "Victorinus (4), ST." Edited by William Smith and Henry Wace. *A Dictionary of Christian Biography, Literature, Sects and Doctrines*. London: John Murray, 1877–1887.

www.ingramcontent.com/pod-product-compliance
Lightning Source LLC
Chambersburg PA
CBHW051423090426
42737CB00014B/2800